SCIENTISTS OF WALES

MONBIWM.
2 6 SEP 2016

21. 10

10. 03. 17.

03. 06

BEAUMARIS

Rhif/No. 3447 6695   Dosb./Class 942.052092

Dylid dychwelyd neu adnewyddu'r eitem erbyn neu cyn y dyddiad a nodir uchod. Oni wneir hyn gellir codi tal.

This book is to be returned or renewed on or before the last date stamped above. Otherwise a charge may be made.

LLT1

SCIENTISTS OF WALES

Series Editor
Gareth Ffowc Roberts
Bangor University

Editorial Panel
Iwan Rhys Morus
Aberystwyth University

John V. Tucker
Swansea University

SCIENTISTS OF WALES

Robert Recorde

TUDOR SCHOLAR
AND MATHEMATICIAN

GORDON ROBERTS

UNIVERSITY OF WALES PRESS
2016

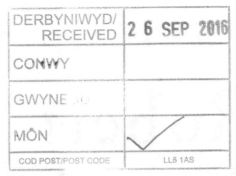

© Gordon Roberts, 2016

All rights reserved. No part of this book may be reproduced in any material form (including photocopying or storing it in any medium by electronic means and whether or not transiently or incidentally to some other use of this publication) without the written permission of the copyright owner except in accordance with the provisions of the Copyright, Designs and Patents Act 1988. Applications for the copyright owner's written permission to reproduce any part of this publication should be addressed to the University of Wales Press, 10 Columbus Walk, Brigantine Place, Cardiff CF10 4UP.

www.uwp.co.uk

British Library Cataloguing-in-Publication Data
A catalogue record for this book is available from the British Library.

ISBN   978-1-78316-829-3 (hardback)
        978-1-78316-854-5 (paperback)
eISBN   978-0-7083-2679-4

The right of Gordon Roberts to be identified as author of this work has been asserted in accordance with sections 77, 78 and 79 of the Copyright, Designs and Patents Act 1988.

Publication of *Robert Recorde: Tudor Scholar and Mathematician* has been made possible with grant assistance from Tenby Museum and Art Gallery.

Typeset by Marie Doherty
Printed by CPI Antony Rowe, Chippenham, Wiltshire.

CONTENTS

Series Editor's Foreword	vii
List of Illustrations	ix
Preface and Acknowledgments	xi
Genealogy of Robert Recorde Physician	xiii
Prologue	1
1 Child of Tenby	7
2 Oxford Scholar	17
3 Cambridge Savant	27
4 Such is Your Authority	39
5 St Paul's Churchyard	53
6 Doctor Recorde	63
7 Antiquarian and Mathematician	75
8 No Mean Divine	87
9 Comptroller of the King's Mints	97
10 The Muscovy Company	107
11 This Talk Delights Me Marvellously	117
12 Pedagogue and Poet	129
13 Surveyor of the Mines and Monies	143
14 Nemesis	153
15 A Heart So Oppressed	165
16 An Unquiet Mind	177
17 One of His Elect in Glory	189

Epilogue	199
Notes and References	207
Select Bibliography	213
Index	217

SERIES EDITOR'S FOREWORD

Wales has a long and important history of contributions to scientific and technological discovery and innovation stretching from the Middle Ages to the present day. From medieval scholars to contemporary scientists and engineers, Welsh individuals have been at the forefront of efforts to understand and control the world around us. For much of Welsh history, science has played a key role in Welsh culture: bards drew on scientific ideas in their poetry; renaissance gentlemen devoted themselves to natural history; the leaders of early Welsh Methodism filled their hymns with scientific references. During the nineteenth century, scientific societies flourished and Wales was transformed by engineering and technology. In the twentieth century the work of Welsh scientists continued to influence developments in their fields.

Much of this exciting and vibrant Welsh scientific history has now disappeared from historical memory. The aim of the Scientists of Wales series is to resurrect the role of science and technology in Welsh history. Its volumes trace the careers and achievements of Welsh investigators, setting their work within their cultural contexts. They demonstrate how scientists and engineers have contributed to the making of modern Wales as well as showing the ways in which Wales has played a crucial role in the emergence of modern science and engineering.

RHAGAIR GOLYGYDD Y GYFRES

O'r Oesoedd Canol hyd heddiw, mae gan Gymru hanes hir a phwysig o gyfrannu at ddarganfyddiadau a menter gwyddonol a thechnolegol. O'r ysgolheigion cynharaf i wyddonwyr a pheirianwyr cyfoes, mae Cymry wedi bod yn flaenllaw yn yr ymdrech i ddeall a rheoli'r byd o'n cwmpas. Mae gwyddoniaeth wedi chwarae rôl allweddol o fewn diwylliant Cymreig am ran helaeth o hanes Cymru: arferai'r beirdd llys dynnu ar syniadau gwyddonol yn eu barddoniaeth; roedd gan wŷr y Dadeni ddiddordeb brwd yn y gwyddorau naturiol; ac roedd emynau arweinwyr cynnar Methodistiaeth Gymreig yn llawn cyfeiriadau gwyddonol. Blodeuodd cymdeithasau gwyddonol yn ystod y bedwaredd ganrif ar bymtheg, a thrawsffurfiwyd Cymru gan beirianneg a thechnoleg. Ac, yn ogystal, bu gwyddonwyr Cymreig yn ddylanwadol mewn sawl maes gwyddonol a thechnolegol yn yr ugeinfed ganrif.

Mae llawer o'r hanes gwyddonol Cymreig cyffrous yma wedi hen ddiflannu. Amcan cyfres Gwyddonwyr Cymru yw i danlinellu cyfraniad gwyddoniaeth a thechnoleg yn hanes Cymru, â'i chyfrolau'n olrhain gyrfaoedd a champau gwyddonwyr Cymreig gan osod eu gwaith yn ei gyd-destun diwylliannol. Trwy ddangos sut y cyfrannodd gwyddonwyr a pheirianwyr at greu'r Gymru fodern, dadlennir hefyd sut y mae Cymru wedi chwarae rhan hanfodol yn natblygiad gwyddoniaeth a pheirianneg fodern.

LIST OF ILLUSTRATIONS

Genealogy of Robert Recorde		xiii
Figure 1	Tenby, Robert Recorde's birthplace	8
Figure 2	Recorde's table of English coins	12
Figure 3	St John's College, Cambridge	28
Figure 4	A urinal, the glass vessel used for the examination of urine	31
Figure 5	Recorde's table for multiplying digits	45
Figure 6	Numbering by the hand	52
Figure 7	The Norman Cathedral of St Paul's	54
Figure 8	Parish of St Katherine Coleman, *circa* 1550	65
Figure 9	Recorde's list of medicines for kidney stones	71
Figure 10	A typical doctor of the Tudor period	76
Figure 11	Recorde's exposition of Pythagoras's Theorem	83
Figure 12	VDMIE, the cryptic slogan of the Protestant Schmalkaldic League	94
Figure 13	Sir William Sharington, Undertreasurer of the Bristol Mint	101
Figure 14	The Castle of Knowledge, atop of which sits Ptolemy	123
Figure 15	Geometry's verdict	137
Figure 16	Ore Stamping Mill and Melting House	146
Figure 17	Sir William Herbert, first Earl of Pembroke	159
Figure 18	Recorde's second part of arithmetic	162
Figure 19	Recorde's table of algebraic symbols	169
Figure 20	Recorde devises the sign for equality	172

Figure 21 Recorde's discussion of 'climates', or time zones   179
Figure 22 The false portrait of Robert Recorde.   203

Picture Credits

Figure 1: View of Tenby by Eric Bradforth, reproduced with the permission of Tenby Museum and Art Gallery.
Figures 2, 4, 5, 6, 9, 10, 11, 12, 14, 15, 18, 19, 20, 21: pages from the extant works of Robert Recorde, reproduced by courtesy of TGR Renascent Books.
Figures 3, 7, 8, 13, 16, 17: public domain images from Wiki Commons.
Figure 22: The Bushell portrait, reproduced with the permission of the Faculty of Mathematics, University of Cambridge.

PREFACE AND ACKNOWLEDGMENTS

Robert Recorde's extant works are printed in Early Modern English, the stage of the English language used from the beginning of the Tudor period until the transition to Modern English during the mid-seventeenth century. The orthography of Early Modern English was similar to that of today, but spelling was unstable. Early printers regarded the letters 'i' and 'j' as interchangeable variations of the same letter, and similarly the letters 'u' and 'v' were not considered distinct or separate. The use of the long form of the letter 's', which modern readers often confuse with the letter 'f', can be particularly troublesome. Printers often made use of abbreviations not found in modern books, and capitalisation, punctuation and hyphenation were usually haphazard. All this can make reading Early Modern English tiresome for anyone unaccustomed to sixteenth-century spellings and typographical conventions.

Accordingly, in the following pages I have transcribed the many quotations from Recorde's texts into modern English, as regards both spelling and punctuation. I have not altered Recorde's phrasing, often peculiar to his times, but have sometimes retained easily understood spellings such as 'hath' instead of 'has', 'doth' instead of 'does', and so on. All quotations are taken from the first editions unless otherwise indicated. Anyone desirous of reading Recorde's original texts can easily do so nowadays by consulting the readily available facsimile editions of his works listed in the bibliography. I have used the same method of transcription when quoting from other sources originating in the same time period as Recorde's texts.

In attempting this biography, which is an avowed work of synthesis, I have drawn on the researches of many dedicated scholars. A book like this could not have been written without their work and I am immensely grateful for the labours and insights of others who have previously traversed this particular historical terrain. I am particularly indebted to the late Professor David Eugene Smith and the late Frances Marguerite Clark for their invaluable efforts in uncovering aspects of Recorde's life and works previously unknown. I have drawn heavily on the unpublished dissertations of Edward Kaplan and Thavit Sukhabanji, to whom I also offer my grateful thanks.

More recently, the scholarly book on Recorde by Jack Williams has revealed new insights and proved to be an inspiration, as has the series of essays by various authors edited into book form by Gareth Roberts and Fenny Smith. To them all I extend sincere thanks. Jo Maddocks of the British Library Rare Books Department, Gill Cannell of the Parker Library, Corpus Christi College, Paul Cox of the National Portrait Gallery and Sue Baldwin† of Tenby Museum and Art Gallery have all been unstinting in their efforts on my behalf. To them I also offer wholehearted thanks. The medievalist Jonathan Mackman has been meticulous in transcribing documents, from both the British Library and the National Archives, written in sixteenth-century handwriting which I found impossible to decipher myself. His skilled and knowledgeable expertise has been of the utmost value. My wife Elizabeth has read and reread my many manuscript drafts and her constructive criticisms have made this a better book. Without her constant support and infectious enthusiasm, not to mention innumerable cups of tea, I doubt whether this book would ever have been completed. The responsibility for errors and omissions remains, of course, with me.

Gordon Roberts
Derby 2015

Genealogy of
Robert Recorde Physician

Genealogy of Robert Recorde Physician
circa 1512-1558

He beareth Sable And Argent Quarterly by the name of Record of Est Well in Cent

1 Roger Recorde
Roger Record Deceased
(possibly in infancy or at a young age)

Joan Recorde
née Joan Ysteven daughter and co-heiress of Thomas Ysteven Gent. of Tenby

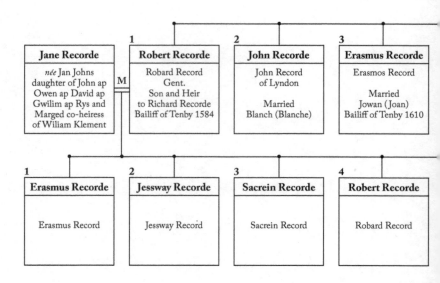

Jane Recorde
née Jan Johns daughter of John ap Owen ap David ap Gwilim ap Rys and Marged co-heiress of Wiliam Klement

M

1 Robert Recorde
Robard Record Gent. Son and Heir to Richard Recorde Bailiff of Tenby 1584

2 John Recorde
John Record of Lyndon
Married Blanch (Blanche)

3 Erasmus Recorde
Erasmos Record
Married Jowan (Joan) Bailiff of Tenby 1610

1 Erasmus Recorde
Erasmus Record

2 Jessway Recorde
Jessway Record

3 Sacrein Recorde
Sacrein Record

4 Robert Recorde
Robard Record

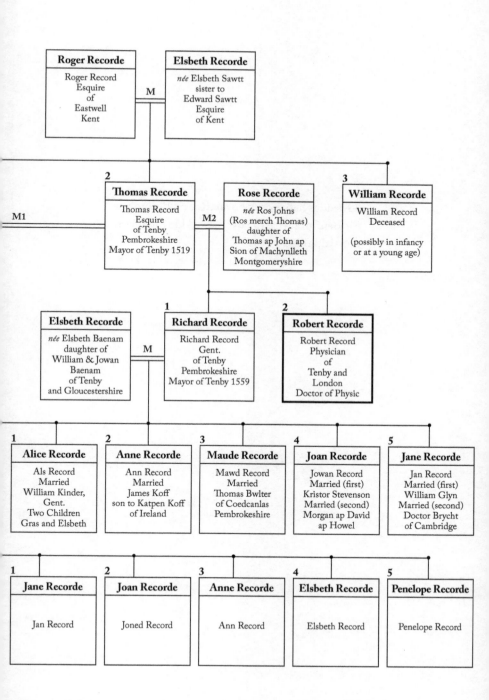

Chart based on the geneology for St John's by Tenby, in *Heraldic Visitations of Wales and Part of the Marches between the years 1586 and 1613* by Lewys Dwnn transcribed and edited with notes by Sir Samuel Rush Meyrick, (2 volumes, Llandovery, William Rees, 1846) Vol 1, pp. 68–69 (alternative spellings of names according to common usage or as written by Lewys Dwnn and recorded by Samuel Meyrick)

'Gweddus hefyd yw cofio am Robert Recorde o Sir Benfro, a wnaeth enw mawr iddo'i hun ym meysydd rhif a mesur.'
—R. T. Jenkins

PROLOGUE

A pen portrait of Robert Recorde, supposedly by a contemporary, states that he was a slight man, brilliant of eye, with a ready smile and a kindly disposition. Unfortunately, this rather pleasing but brief sketch has no attribution and is unverifiable with regard to his features. The same tenuous source says that he loved nothing better than debate, that he was deeply skilled in rhetoric, philosophy, literature, history, cosmography, physic, mineralogy and every branch of natural history. He is said to have preached well and to have had a sound knowledge of the law.[1] All this accords with everything known about this distinguished scholar, although it brings us no nearer to envisioning his appearance.

W. F. Sedgwick, in an 1896 entry in the *Dictionary of National Biography*, unequivocally stated that the only known portraits of Recorde were woodcuts in two of his books. This, however, proved to be no less a chimera than the unattested word picture described above. In a 1921 edition of the *American Mathematical Monthly*, the editor said that examination of copies of Recorde's works held in the Library of Congress and the Surgeon General's Library in Washington, and copies in the British Museum Library and in the Bodleian Library at Oxford, showed that no such portraits existed. He confirmed that the title page of a medical treatise by Recorde did contain a woodcut depicting a typical doctor of the period, and his text on geometry had a scholar at his desk within an illustrated letter 'G' (part of the word Geometry). He also pointed out that it was not uncommon for woodcuts in early European books to be regarded as portraits. However, Mr. A. W. Pollard of the Department of Printed Books in the British Museum saw no reason whatsoever to treat these rather crude images as portraits of Recorde. So faded away any chance we might have had of catching even a glimpse of his physiognomy, or so it seemed.

Then came a remarkable development. In 1923 Professor David Eugene Smith, a distinguished historian of mathematics, announced the discovery of Recorde's portrait in oils, declaring that 'as to the authenticity of the painting there can be no question'. He published the picture in his two-volume *History of Mathematics*, but he was sadly mistaken as to its veracity. Although long believed to be genuine, the painting is now discredited. The history of this portrait, purporting to be Recorde's likeness, is an interesting one and will be discussed in the epilogue to this book. Suffice here to say that, sadly and contrary to what is popularly believed, Recorde has no known portrait and we have no idea what he looked like.

Nor is it easy to form a mental picture of the whole man. Snippets of his life story, written by commentators and chroniclers in the years after his death, and continuing up to the present day, lie scattered and unanalysed. Probably no other important historical figure of the Tudor period has been so neglected by modern scholars, and heretofore no attempt has been made to gather up these bits and pieces into a coherent and chronological whole. As pointed out by Thavit Sukhabanji, this may be because Recorde is perceived primarily as a populariser of mathematics, rather than as a great theoriser who made significant advances in the development of the mathematical sciences.[2] Accordingly, historians of the Tudor age have shunned him, preferring to leave anything remotely connected with mathematics to be written about by mathematicians. Conversely, historians of mathematics have tended to show little interest in the many other facets of Recorde's eventful and ultimately tragic life.

Recorde was a Tudor scholar and Renaissance scientist of the first order, and in his own time his reputation for learning was second to none. His influence on his contemporaries and the following generations of Elizabethans was immense. He was at once a physician, a teacher, an accomplished mathematician and a savant whose breadth of knowledge in the sciences was astounding. He was a great humanist scholar who embraced and expounded a wide array of philosophical ideals. He championed the Protestant Reformation and flourished, sometimes precariously, during the volatile and deadly dangerous times of religious upheaval. Eventually he became entangled in the murky world of Tudor politics and, out of his depth in double-dealing and intrigue, ultimately succumbed to its dangers.

The Catholic scholar John Pitts, among notices of many contemporary authors in his *De Illustribus Angliæ Scriptoribus*, published in Paris in 1619, has left us a valuable characterisation of Recorde. Writing in Latin, this is what Pitts said about him:

> A happy man of genius, and famous for teaching complex subjects. A polished and accurate writer, highly skilled in all the liberal arts and the mathematical sciences. He was a most famous philosopher. He contemplated the motions of the heavenly bodies, and achieved considerable expertise in astronomy. He probed the secrets of natural philosophy, of plants, herbs, roots, of the elements, and examined with curiosity the strengths and virtues of the metals. It may be said with good reason, that he climbed into the heavens and penetrated the bowels of the earth.

Robert Recorde was a Welshman, long domiciled in England but born in Pembrokeshire, in the south-west corner of Wales. The family name, however, is not Welsh but Norman-French in origin. Norman personal names such as Richward, Richold, Ricard, and the more popular Richard, were first introduced into England after the Conquest of 1066, and these spellings gradually fused into the surname Rikeward, which eventually metamorphosed into Recorde. Today, people with this surname usually spell it without the final 'e'. A Norman ancestry perhaps explains the family liking for male names such as Thomas, John and William, and especially names which give a double 'R' alliteration, such as Roger, Richard and Robert Recorde, all of which reoccur in later generations.

Recorde is reputed to have been born of a good family, 'genteel' in the words of Anthony à Wood, which suggests that they were at one time regarded as gentry.[3] This is borne out by the possibility that the family was armigerous, its arms awarded at some unknown time in the distant Norman past and blazoned as quarterly sable and argent (or, less formally, a shield quartered black and silver). The source of this supposition, together with all other genealogical information about the Recorde family, originates in a visitation to Wales by Lewys Dwnn, deputy herald at arms, in 1597. Heraldic visitations were tours of inspection through England, Wales and Ireland, in order to regulate and register the coats of arms of nobility and gentry, and to record pedigrees.[4]

Recorde's paternal grandfather was named Roger, and in the last quarter of the fifteenth century he made what must have been an extraordinary decision for his time. In an age when people instinctively remained close to kith and kin and few travelled much further than the nearest market town, he decided to uproot his family, leave his native Kent, and emigrate to what most English people would have regarded as a foreign land. His country of choice was Wales, a destination even more remarkable because at that time the English regarded the principality as a lawless region, where the king's writ did not run. Its hills and deep valleys were popularly supposed to be infested with outlaws, remnants of the bands who had supported the rebellion of Owain Glyndŵr half a century earlier. At this time also, local rivalries between Welsh landowners often broke out into violent quarrels, looked upon askance by the few Englishmen who thought anything at all about this alien and misunderstood land.[5]

It is possible that as a merchant Roger Recorde had cause to travel more than most. Residing in the small hamlet of Eastwell, close to the market town of Ashford and not far from Canterbury, he was probably familiar with the Dover road and the route from London followed by pilgrims visiting the shrine of St Thomas à Becket. In the course of his business he may even have travelled as far as London himself, as well as visiting the ports on the Channel coast. This would explain why he was apparently undaunted by the many miles which lay between Eastwell and Wales, and the difficulties of such a journey at a time when there were no maps and the highways were little more than rutted or muddy cart tracks and drovers' roads.

Roger married Elsbeth Sawtt, described as sister to Edward Sawtt, which probably meant that her parents were deceased at the time of her marriage. The ceremony possibly took place in the medieval church of St Mary the Virgin, the remains of which today stand in ruins in Eastwell Park, scheduled as an ancient monument. In due course Elsbeth presented Roger with their first son, whom they named Roger after his father. They would eventually have two more sons, Thomas and William, who were probably not born until the family was safely domiciled in Wales.

Why and exactly when Roger Recorde decided to emigrate to Wales remains an open question. It is plausible that his thoughts first turned to the principality because of momentous events that occurred around the time he and Elsbeth must have begun thinking about embarking on their adventurous journey. In 1485 Henry Tudor, Earl of Richmond, accompanied by his uncle Jasper Tudor and a small force of French and Scottish soldiers, returned from a long exile in France with the intention of seizing the throne of England from Richard III. They landed in Mill Bay on the coast of Pembrokeshire and, probably because of his Welsh birth in nearby Pembroke Castle and his descent through his father from Rhys ap Gruffydd, Henry quickly amassed an army of five thousand Welshmen to add to the soldiers he had brought with him from France. On 22 August he engaged Richard's army in the battle of Bosworth Field, and Richard was killed in the fighting. Hurrying to London, Henry cemented his claim to the throne by marrying Elizabeth of York, daughter of Edward IV, and in the process founded the Tudor dynasty under which Roger and Elsbeth's grandson Robert Recorde was to live out his life.

Englishmen, now subject to a Welsh monarch, surely and quickly perceived Wales from a new and more enlightened viewpoint. The county of Pembrokeshire – Sir Benfro in Welsh – must especially have been on many people's minds, excitingly associated as it was with their new king. Roger may also have been influenced to make Pembrokeshire his destination in the knowledge that this corner of Wales had been English in language and culture for many centuries, despite its remoteness from the English border. Called 'Little England beyond Wales' and referred to in Welsh as 'Sir Benfro Saesneg', meaning English Pembrokeshire, one can imagine Roger assuring Elsbeth that it wouldn't be so different after all from their old home in Kent. The unmarked but traditional boundary between English and Welsh speakers (with Welsh to the north and east, and English to the south and west) stretches eastwards from St Bride's Bay in the west of Pembrokeshire, until it meets the river Taff north of Laugharne in Carmarthenshire. George Owen, writing his *Description of Pembrokeshire* at the end of the Tudor period, said: 'you shall find in one parish a pathway parting the Welsh and English, and the one side speak all English, the other all Welsh'.

The journey to Wales by Roger and Elsbeth would have been on foot or horseback, child in arms, and, as was customary with travellers, their stock of money sewn into their clothing. From time immemorial the standard aid for travellers was the itinerary, a simple written list tabulating the points of departure and destination, with intervening places and possibly the distances between them included. They would have known where they were headed day by day, and knew also that so long as they kept to the highway they could always ask for directions. So without maps, seeking overnight shelter from friendly cottagers or the hospitality of monastic houses, and as a last resort availing themselves of expensive accommodation at wayside inns, waiting often for fellow travellers to provide safety in numbers, they might have headed for Gloucester and there crossed by the bridge over the Severn into Wales. If so, an equally long and arduous journey along the estuary coast, over tracks unbelievably worse even than those of England, would have eventually brought them within sight of the walled town and port of Tenby, their final destination. There was, however, an alternative and easier route, which might have been suggested to Roger by the example of Margaret Beaufort, Henry Tudor's mother. It was well known that she and her entourage had gone by ship from Bristol across the Severn estuary to Chepstow, and the couple could possibly have followed her lead. Their first sight of Tenby, therefore, may well have been from the deck of a sailing vessel far out in the Bristol Channel.

1

CHILD OF TENBY

Whether Roger and Elsbeth Recorde stepped ashore in the small harbour, or entered by road through one of the gatehouses set in the town walls, they would have been reassured by Tenby's strong defences. Although the Norman castle was in disrepair and largely abandoned as a defensive fortification, Jasper Tudor had concluded an agreement with the town's merchants to split the costs for refurbishing Tenby's ramparts. The dry ditch along the outside of the town walls was widened to 30 feet, the walls were heightened and a second tier of arrow slits were pierced above a new parapet walk. Additional turret towers were added to the ends of the walls where they abutted the cliff edges. It must have been clear to Roger that Tenby was a place where a newcomer, a merchant, could safely establish himself in business.

Tenby lies on the west side of Carmarthen Bay, looking out over the estuary towards St George's Channel, the Irish Sea and the distant Atlantic. At the time of Roger and Elsbeth's arrival it was a thriving port with a lucrative sea trade across to Bristol and the south and west coasts of England, as well as further afield to Dublin, France, Spain and Portugal. Originally a fishing port anciently named in Welsh Dynbech-y-pyscoed, which is today spelt Dinbych-y-pysgod, it owed its rise to the settlement of Flemings who established a woollen trade here in the reign of Henry I. Roger Recorde may have sensed mercantile opportunities in the wool markets, and this could have been the attraction which lured him to the town from Kent. Tenby was one of Wales's busiest ports, with a bustling trade on the harbour side and in the nearby streets, which were lined with the houses of merchants. We do not know

FIGURE 1 Tenby, Robert Recorde's birthplace

This view of the walled town of Tenby, by Eric Bradforth, is based on a survey of 1586. The town is dominated by the church of St Mary the Virgin, reflecting the prosperity of the medieval port. The small seaman's chapel of St Julian can be seen on the end of the quay, near the harbour entrance.

where Roger and Elsbeth made their home in Tenby, but in due course Thomas and William, brothers to Roger, were born here. Sadly Roger and William were to die young, perhaps in infancy, leaving Thomas as the sole survivor to eventually inherit the family's mercantile business.

Nothing is known of Thomas's early years, but inevitably the day arrived when he sought a bride. He did not look beyond the local gentry, making a good marriage with Joan, the daughter and co-heiress of Thomas Ysteven of Tenby. Ysteven was a man of some importance, a bailiff in 1462 and three times mayor in later years. How long this marriage lasted is not known, but Joan died without issue and Thomas became a widower. He won himself a second bride, Ros, from the town of Machynlleth in Montgomeryshire, probably while travelling far afield from Tenby dealing in wool, skins, hides and cloth, which were the staple produce of the Welsh hinterland beyond Little England.

Ros was the daughter of Thomas ap John ap Sion, the father's patronymic name suggesting a long Welsh ancestry. It was the custom in Wales for a person's baptismal name to be linked by *ap* (son of) or *merch* (daughter of) to the father's baptismal name, down sometimes to the seventh generation. It is thought that the patronymic system arose from early Welsh law, which made it essential to know how people were descended from an ancestor. So Ros, after her father's example, would have given her name as Ros merch Thomas ap John. However, around the time that Thomas Recorde met Ros, the Welsh patronymic naming system was slowly yielding to the English system of fixed family names. The most common surnames in modern Wales result from adding an 's' to the end of the name, so that Ros may well have been introduced to Thomas Recorde in the English fashion as Ros Johns. This digression into naming conventions serves to show that Robert Recorde's mother, for so was Ros to become, was *not* named Jones as is so often erroneously stated. The Welsh spelling of her first name as Ros, also variously spelt as Rhos or Rhosyn, can be translated into English as Rose, a name with which she is often credited by English writers.

It is not known where the marriage of Thomas Recorde and Ros Johns took place, but the couple made their home in Tenby. Ros may not have felt immediately comfortable in Little England, a long way from the Welshry of her family. George Owen said of the English side of the language divide, that they:

> keep their language among themselves without receiving the Welsh speech or learning any part thereof, and hold themselves so close to the same that to this day they wonder at a Welshman coming among them, the one neighbour saying to the other 'Look there goeth a Welshman'.

Nevertheless, Ros must have reconciled herself to living in the Englishry with her husband and in due course Richard, the first of their two sons, was born. Robert followed, his birth usually ascribed to the year 1510. However, this is certainly too early and 1512 is a much more likely date, as will be adduced as this history unfolds.

Despite being born in Little England, and although only of the second generation on his father's side, Robert was of course entitled to the appellation 'Welshman', not least because, in the words of Anthony à Wood, he 'received his first breath among the Cambrians'.[1] However, it was on his mother's side that Robert could lay claim to a long Welsh heritage. It is pleasing to think that she may have taught her sons something of the Welsh language, and that the lilt of spoken Welsh was not entirely absent from their home. The concept of childhood as we know it today scarcely existed in Tudor times, and young children were thought of as immature adults without the strength as yet to undertake hard manual work or the mental capacity required for reasoned thinking. Nevertheless, Richard and Robert, dressed in smaller versions of adult clothes and encouraged to help around the house from a very early age, would have gradually became familiar with their father's business.

We do not know where the family home in Tenby was situated, but as the house of a middle-class merchant, it would have had a large ground floor room opening out on to the street. This was where Thomas did his trading, the room serving as shop, office and warehouse, the goods for sale spilling out into the street, the purchased stock safely stored at the back.[2] Near the door would have been the counting table, its surface inscribed with parallel lines on which were placed small, round counters. The lines represented units, tens, hundreds, thousands and so on, and a skilled merchant could, by rapidly moving the counters to the different lines, quickly calculate a bill or payment. Once an account was finalised it was recorded in ledgers using Roman numerals. Weights, measures and coinage were all extremely complicated at this time and required a good head for arithmetic.

Methods of calculating using newly introduced Arabic numerals, written with a pen on paper, were hardly known to anyone during Robert's childhood. Nor did most people have any idea of how to calculate using the much older method of moving counters on a board. It was regarded as a black art, its merchant practitioners looked upon with suspicion as magicians, so that magic became synonymous with mathematics in the minds of many of the more credulous. After all, the methods used to cast an account seemed little different from those used to cast a spell. Robert could hardly have known that a major part of his life's work would consist of dispelling these superstitious fears, and teaching people the simple means by which they could carry out calculations for themselves in an increasingly mercantile world.

Tenby's market was the oldest in Pembrokeshire, its initial charter being granted in 1290, and no doubt Robert would have become well acquainted with the busy stalls situated in the town centre. Fascinated by the variety of goods for sale, we might suppose that he began to develop his lively and enquiring mind as he explored the market and its environs in the company of his elder brother. Watching the business of buying, selling and bartering merchandise, all of which had its own peculiar system of weights and measures, would soon have made him realise that a sound knowledge of calculation was an essential element of daily life, at least for the mercantile classes. In a future time he would write a book containing an explanation in simple terms of the relationships between the different weights and measures, and the many different coins of the realm that were then used in exchange.[3]

Probably the most exciting place for the brothers to explore, without going outside the town walls, was the harbour. Inside the curving quay on the north side, near a narrow entrance made narrower by a breakwater of stones stretching out from under the cliff face, they would have seen a forest of masts and rigging. Prominent among small coastal vessels and trows from the higher reaches of the Severn, they would have watched caravels from Portugal and carracks from Spain discharging and loading their cargoes. These were ocean-going ships, large enough to be stable in heavy seas, and roomy enough to carry provisions for long voyages. We can imagine Robert's excitement when he learnt that these ships were of the very type in which daring Portuguese and Spanish seamen were even then probing the Atlantic coasts of Africa and crossing the seemingly limitless ocean to the Americas, exploring and mapping the known limits of the world.

On the harbour side, awaiting shipment to England upstream on the Severn or via the south coast ports, or export to Ireland, Brittany or the Iberian Peninsula, would have been barrels of grain, wooden tubs of butter and cheese, bales of raw wool and woven woollen stuffs, cured

> ### Reduction.
>
> men coynes, weyghtes, meafures, & fuche other, I haue prepared here a brefe table, which fhall fuffyce to you at this tyme, tyll hereafter at more cōuenient oportunite, I maye enftructe you more exactely in the fame.
>
> ⁋ A table for Englyfh coynes. *Englifh coynes.*
>
> | | |
> |---|---|
> | ⁋ A fouerayne. | A quarter noble. |
> | Halfe a fouerayne. | A crowne. |
> | A royall. | Halfe a crowne, |
> | Halfe a royall. | A crowne. |
> | A quarter royall. | A grote. |
> | An olde noble. | A harpe grote. |
> | Halfe an olde nob. | A peny of 2 penes |
> | An angell. | A dandypratte. |
> | Halfe an angell. | A penny. |
> | A george noble. | An halfe penny. |
> | Halfe george nob. | A farthynge. |
>
> ⁋ The valowe of Englyfhe coynes.
>
> A Souerayne is ẏ greateſt Englyfh coyne, and contayneth 2 Royalles or 3 Angelles, eyther 9 halfe crownes, or 4 crownes and an halfe, that is to faye 22 s̷. 6 d̷. *The valewe of Englyſhe coynes.*
>
> Halfe

FIGURE 2 Recorde's table of English coins

The coins listed are: a Sovereign, half a Sovereign, a Royal, half a Royal, a Quarter Royal, an old Noble, half an old Noble, an Angel, half an Angel, a George Noble, half a George Noble, a Quarter Noble, a Crown [5s], half a Crown, a Crown [4s 6d], a Groat, a Harp Groat, a Penny of two Pennies, a Dandyprat, a Penny, half a Penny, a Farthing. (*The Grounde of Artes* (1543), p. 88.)

skins and hides and even small amounts of coal and lead. Robert would no doubt have been intrigued by negotiations carried out on the harbour wall between ship's captains and town merchants who did not speak each other's language. Buying and selling prices for inward and outward bound cargoes were agreed by age-old hand signs, rapidly passed back and forth by men long accustomed to understand them. Robert would later explain and illustrate this method of signing in one of his books, calling it 'the art of numbering by the hand'.[4] He must have watched laden ships leave the harbour and, as he grew older, observed that hulls on the distant horizon disappeared from his sight long before the top masts. Here, before his eyes, was proof that the world really was a globe, for only curvature of the earth's surface could account for this phenomenon. He would write about this too, in due course.

Robert would also have been familiar with the sight of seamen slipping in and out of the small stone chapel at the end of the quay, to offer thanks for a voyage just completed or to pray for a safe passage about to begin. The entire western world at this time was under the religious domination of the Roman Catholic Church, so mariners of all nationalities would have felt at home in this tiny building dedicated to St Julian.[5] Robert's enquiring mind might have prompted him to peep in at the doorway, but he would have attended services with the other members of his family at the church of St Mary the Virgin, high on the cliff top in the centre of the town. Every Sunday the parishioners of Tenby, almost without exception, the Recorde family among them, would have made their way to church as people had done for generations past. The central nave would have been overcrowded, the crush tolerable because pews were not yet in fashion and the custom was for people to stand or kneel in the empty space.

A rood screen, probably solid at the bottom but with narrow arches above, would have separated the nave from the chancel, making it difficult for the parishioners to see the high altar raised against the east wall. The main service of the day was the high mass, for which the clergy donned elaborate vestments. Once robed, they consecrated holy water and processed around the church, accompanied by assistants carrying the processional cross, an incense burner and the sacring bell, sprinkling the holy water on the congregation and the altars in the side chapels as they went. Passing through the rood screen into the chancel, the service proper would begin at the altar steps. The priest would place a wheaten wafer on the communion plate and pour wine into a communion chalice, all the while reciting elaborate texts and following rubrics that specified every movement and gesture. The laity, Robert among them and all squashed together in the nave, could undoubtedly smell the incense, but the intervening screen meant they could see very little of the opening rituals. Few of the congregation would have understood the priest's oratory as he deliberately kept his voice low and, in any case, everything was recited in Latin. While the clergy participated in the liturgy, the parishioners were expected to engage in private devotions.

Up until this point the parish mass was a ceremony that the laity observed as best they could, rather than participated in. However, the priest, speaking in English, would then call on the people to pray for the pope and the clergy, for the king, and for those in special need, such as pregnant women and the sick, and finally for the dead, especially recently deceased parishioners. The priest would then return to the chancel and the ringing of the sacring bell announced the climax of the mass, when he would repeat the words of Christ at the Last Supper – *'hoc est enim corpus meum'* – 'for this is my body' – as he elevated the newly consecrated wafer, the host, high above his head. The chalice too was raised on high with the words *'hoc est enim calyx sanguinis mei'* – 'for this is the cup of my blood'. Only the priests afterwards consumed both wafer and wine in what was called 'communion in both kinds'. For the laity communion took the form of kissing the paxbred, a plate adorned with a sacred picture, which the priest had kissed immediately after kissing the lip of the chalice and the cloth on which the host rested.

We can guess from Robert's lifelong and pious devotion to the church that he must have found these wondrous and magical happenings in his early life both inspiring and indeed spiritually essential. He could not have had any idea that within his lifetime the Roman Catholic Church, with all its centuries-old ritual and splendid ceremonies, would collapse like a house of straw under the ferocious onslaught of a competing doctrine. Nor could he have foreseen that during this process his life would be irrevocably changed, and sometimes even placed in jeopardy.

As they grew older Robert and his brother would have been expected to learn the Catholic creed and catechism and to repeat them in Latin, if only in parrot fashion. This would be Robert's first exposure to the language of scholarship, the *lingua franca* of educated people throughout Europe, the tongue in which he would later become dazzlingly adept.

In 1519, when Robert was about seven years old, his father was elected Mayor of Tenby, which undoubtedly led to an increase in the family's status within the town. Thomas shared the mayoralty for part of the year with a William Thomas, but the reason for this is not known.[6] As the sons of such a prominent and well-to-do citizen, the two boys, perhaps having learnt to read and write at their mother's knee, may have been educated by a paid schoolmaster or mistress, although it is equally feasible that they were educated by the clergy.

St Mary's Church was extensively rebuilt during the fifteenth century and a new door with a large cruciform porch was inserted in the west wall. Also erected nearby was a so-called college, possibly for the use of the chantry priests serving the three chantry chapels within the church. Both college and west porch are now demolished, but there is evidence that the porch was used as a schoolroom. In 1657–8, payments were recorded for the repair of the church windows and those of the 'scholehouse', and it is entirely possible that Robert received his early education in one or other, or both, of these buildings. The provision of teaching was one of the many duties assumed by the church, and the education of the young – at any rate the boys – was an important part of the Christian ideal. The chantry priests were not only enjoined to sing mass for the souls of the chapel founders, but also, in their considerable spare time, to instruct the young. In addition to developing the vernacular reading and writing skills of their pupils, the priests would teach them the elements of chanting in Latin so that they could assist in the celebration of church services. Robert would have been taught no mathematics here though, as even the simplest forms of arithmetic were beyond the abilities of most of the clergy who were otherwise able and willing to teach.

Simple ABCs were available at the time Robert began his education, and the early printers were already supplying an increasing demand for more ambitious works intended for young readers. Although the bible was not yet printed in English, reading and handwriting may have been practised and improved by reading sermons and other homiletic works and then copying them out on to slates. Schoolboys could also have read jest books, such as the *Demands Joyous*, or books of carols and a variety of romances, assuming their religious mentors were not so strict as to forbid them. From what we know of Robert's later achievements, we can assume that by the time he reached adolescence he was a bright and intelligent boy, with a depth of knowledge perhaps unusual for his tender years. The mercantile calculations carried out in his father's shop may have given him his taste for mathematics. Tenby harbour had provided him with knowledge of the sea and ships and rudimentary notions of navigation. Stacks of lead and coal awaiting shipment might have aroused his interest in the origin of minerals. The many languages overheard in the harbour, together with a mix of Welsh and church Latin, probably attuned his ear to the sound of other tongues. These things all played a part in his later life, and it is probable that as a boy he displayed such lively precocity that to go on to university seemed a natural progression.

We do not know who first suggested that he should go to Oxford. It may have been his parents, recognising in him abilities they wished to nurture. The chantry priests might have anticipated the possibility of him entering into holy orders. It may have been his own longing for erudition, his lifelong passion for learning already welling in his breast. Whatever the circumstances, at about the age of thirteen, Robert said goodbye to his childhood home and the town of Tenby, and turned his face towards Oxford.

2

OXFORD SCHOLAR

Recorde entered Oxford as a barely adolescent youth in 1525.[1] If we assume he followed customary practice and began his university studies at the age of thirteen, we have the first circumstantial evidence that his date of birth was in fact 1512, and not 1510 as generally accepted. It is possible he made the journey from his home to Oxford under the care of a 'fetcher' or 'bringer of scholars'. These were trusted servants of the university, protected by special enactments, whose business it was to collect aspiring students from their districts and conduct them, in groups of twenty to thirty, to the university towns. It would have been a convenient way for Recorde's father to see his son safely delivered to Oxford, without the necessity of a return journey by himself or his servants.

As the son of a merchant, it is unlikely that Recorde would have been in possession of a scholarship. On arrival therefore, along with fellow newcomers, he would have been speedily enrolled under a tutor, usually a Master of Arts who kept a hostel, that is, a private boarding house licensed by the university. His first lessons in self-reliance would have been swift, as the university in its corporate capacity did not concern itself overmuch about the instruction or discipline of its youngest members. Life was hard and the times were rough and if any student came to grief, no one in authority was likely to care very much. Students would have quickly realised that if they found themselves in a hostel which was badly run with regard to discipline, sleeping quarters and the provision of food, they would soon be in very sorry straits indeed, without any means of appeal to the university for betterment of their conditions.

Once ensconced within the university Recorde would have commenced four years of study of the trivium, which consisted of courses in Latin grammar, logic and rhetoric. Few students who came up to Oxford knew anything beyond the basic elements of Latin, and they would have had first to unlearn any so-called church Latin, with its Italianate pronunciations. The first year then was spent in learning to read, write and speak classical Latin, the language of scholarship throughout Europe. This entailed studying grammar laid down by Priscian, but also the lighter reading of Terence, Virgil and Ovid, supplied in the form of texts probably copied out by the students themselves from a single exemplar.

The next two years were devoted to logic, the texts again being copied out by each student for his personal use. Essential to their studies would have been tracts on logic by Petrus Hispanus (Peter of Spain), Peter Lombard and John Duns Scotus. The standard version of Scotus's text used at Oxford, which Recorde would have read, was the *Opus Oxoniense*, an amended version of the lectures he gave as a bachelor at Oxford and which he revised about 1300. The venerable age of these tracts by the time Recorde studied them is an indication of the lamentable stagnation of university teaching throughout the later medieval period, a situation not modified in any meaningful way until after he quit academic life.

Nevertheless, we can imagine Recorde's fascination as he contemplated the location of angels as a starting point for a complex discussion on the logic of continuous motion, and whether it was possible for the same thing to be in two places at the same time. He would also have studied logical questions such as how angels can be different from one another, given that they have no material bodies, in order to investigate the difficult question of individuation in general. The last year of the trivium was given over to rhetoric. This involved the study of selected parts of Aristotelian philosophy, the concluding texts of which were intended to take students on to the next phase of their studies.

Recorde would have been taught in 'schools', buildings which were hired by the lecturers and by which means the university earned a considerable part of its scanty income. It is likely that during lectures

Recorde and his fellow students would have been expected to sit on straw scattered on the floor. Both Oxford and Cambridge had adopted and long perpetuated a statute instituted in 1366 by Pope Urban V for the University of Paris which, in an effort to prevent ennui among students, forbade them the use of benches or stools in lecture rooms.

By the fourth year of his studies Recorde was expected to have read all the subjects of the trivium. During this final year students were required to dispute publically in the schools four times, twice as a respondent to defend a thesis that they asserted, and twice as an opponent to attack those asserted by others. These disputations were known as acts and the subject matter was usually some scholastic question, or a matter of religion concerning the interpretation of biblical texts. The performance or 'keeping of his acts' would have brought Recorde to the notice of the university senate for the first time, as his moral character and academical standing was investigated. If nothing was reported against him, he would then have been allowed to present himself for examination by the proctors and regents in the Faculty of Arts. This was a severe ordeal for all aspiring degree students, as they were expected to defend a thesis against the most practised logicians in the university. It is said that stupid men propounded some irrefutable truism, but ambitious students courted attack by affirming some paradox. Recorde would surely have belonged to this latter group.

A candidate was never failed or rejected at this examination, but academic reputation or contempt followed the popular verdict as to how he had acquitted himself. Accordingly, it was desirable to have a friendly and supportive audience on these occasions, and the practice grew where not only did the candidate's friends attend, but every passer-by was also forcibly persuaded into joining the band of supporters. This practice became such a nuisance that in Recorde's time the university passed a statute condemning it under penalty of excommunication and imprisonment. Having survived the ordeal of examination, Recorde would have been considered an incepting bachelor. This was not the degree itself, but it marked his transition from schoolboy to the life and studies of an undergraduate, and as such was accompanied by an appropriate ceremony.

At about nine o'clock on the appointed morning, the beadles, each carrying his silver staff of office and accompanied by the sound of pealing bells, gathered the hooded and gowned masters, bachelors, scholars and inceptors and led them in procession to assembly. Once all the company was present, a beadle would direct the youngest inceptor to come bareheaded before the dean to kneel and be commended on his successful studies. Each of the inceptors would follow in turn, and the beadle would pull the hood of each candidate over his face, so that blushes raised by their modesty could not be seen. The dean would then ask a question and expect to receive a satisfactory answer from each candidate. This was purely ceremonial and the dean was not supposed to entrap candidates into argument by asking questions too difficult to answer readily. If he did, the regulations ordered the beadle to 'knock him out', that is, to rap so loudly on the floor with his staff that the noise drowned his voice and nothing else could be heard above the resulting tumult.

For the rest of the year the newly incepted bachelors were expected to spend every afternoon in the schools. Here they were obliged to dispute with any regent or proctor who cared to come and test their abilities. In addition they were required to preside at least nine times over the disputations of those juniors who were still studying the trivium, criticise their arguments and sum up the whole discussion. A master was present to keep order on these occasions, but this was often quite beyond his powers. Drinking, wrestling, cock-fighting and archery with the crossbow were common amusements on these afternoons, and the school was a rough one. Although regarded as a great opportunity for distinction, many of the bachelors preferred to carry out these presiding duties by proxy, which was permitted.

On 16 February 1530, Recorde was finally admitted Bachelor of Arts.[2] It was common practice for bachelors to leave the university at this point and seek their future livelihoods elsewhere. Those who chose to stay, as Recorde did, were expected to devote the next three years to the study of the quadrivium. This consisted of arithmetic, geometry, music theory and astronomy, which, together with the three subjects of the trivium, comprised the seven liberal arts. In view of his later achievements in the sciences, we may be sure that Recorde diligently studied all these subjects, although his lecturers were likely to have been second-rate men. Oxford placed little emphasis on the quadrivium at this time and formal teaching probably went no further than the traditional and narrow limits brought forward from the medieval period.

In the following year (1531) Recorde was elected a fellow of the College of All Souls of the Faithful Departed.[3] As a fellow, Recorde would have been pleased to receive an annual livery of cloth, free quarters and commons (an Oxford term for allowances and meals shared in common), and battels (another Oxford term meaning an account with the college for extra kitchen and buttery expenses, clothing and books). All Souls was founded to produce learned clerics to serve Church and State, and also to be a chantry where the fellows would pray for the souls of the deceased. The fellows were accordingly under a double obligation, to take holy orders and to engage in higher studies, either for a doctorate of theology or in the two laws, canon and civil. The statutes do not refer to medicine, the third higher faculty, but it was not unusual for a number of fellows to proceed to medical doctorates. Recorde was probably earning small sums of money from the cursory lectures he was obliged to give at this time, but he needed to look to the future. On his entry to All Souls he decided to study medicine and make it his future career.

Recorde was evidently both diligent and successful in his medical studies because in 1533 he was awarded an Oxford licence to practise medicine.[4] This did not make him a fully qualified doctor, but it did allow him to gain medical experience by diagnosing and treating the illnesses of real patients, so long as the maladies he treated were within his competence. To obtain his licentiate, which was a degree somewhere between a bachelor and a master, Recorde was first required to attend two anatomies. This involved being present in an anatomical theatre as a professor of medicine, seated above a cadaver, gave lectures on human physiology and anatomy whilst directing an assistant first to open the torso and point out the heart, lungs, kidneys, viscera and the main arteries, before turning to the muscles controlling legs, arms, fingers and the neck. Following this Recorde had to demonstrate at least three cures, which was probably not difficult because illnesses serious enough to warrant a doctor's intervention usually had two outcomes; the patient either recovered or did not. Those who died were not in a position to say anything derogatory about their treatment, whilst those who regained their health were all too ready to attribute their recovery to the skill of the doctor.[5]

At the end of the three-year quadrivium there was no public test of knowledge and all that was necessary to proceed to the degree of Master of Arts was to make a simple declaration that the course had been studied in its entirety. Recorde probably never made that declaration as, according to Anthony à Wood, whether he took 'the magisterial degree the public registers show not'.[6] These were times of great indiscipline within All Souls, and failure to take degrees and enter holy orders, of which infraction Recorde seems guilty, was of serious concern to the Church. Something had obviously gone very wrong for Recorde and we need to retrace our steps a few years in order to see what that might have been.

John Wycliffe, in his time a prominent theologian, had been dismissed from Oxford in 1381 for criticising the Church. He attracted followers, usually without any academic learning and so unable to read Latin, who supported his translation of the bible into English. These followers were called Lollards, a popular derogatory nickname, and believing the Catholic Church to be corrupted in many ways, they looked to Scripture as the basis for their religious ideas. They denounced doctrines such as transubstantiation, exorcism, pilgrimages and blessings, arguing that emphasis should be on the bible rather than church ritual. By the time of Recorde's early years at Oxford, Lollards were attacking the provision of chantries and prayers for the dead, the excesses of church artwork and worshipping of icons, and they proclaimed that confession to a priest was unnecessary, since only God could forgive sins. As a pious Catholic, Recorde would no doubt have been astonished at these assaults on the beliefs and practices he had absorbed since infancy from the priests of St Mary's in Tenby.

At about the time of his election to All Souls, Recorde became acquainted with John Robyns. Although Robyns was ten or twelve years older than Recorde, a friendship soon developed between the two men. Robyns, like Recorde a fellow of All Souls, served as arts bursar to the college in 1529–30, and was ordained in 1532. In the same year he became a canon of Henry VIII's College, Oxford, and a chaplain to Henry himself. According to Anthony à Wood:

> such was his vigorous genie, that by the force thereof, being conducted to the pleasant studies of mathematics and astrology, he made so great progress in them, that he became the ablest person in his time for those studies, not excepting his friend Recorde, whose learning was more general.[7]

By this time Recorde was well on the way to acquiring a reputation for learning and erudition in the sciences himself, and it is likely that Robyns had a great influence on the development of his thinking. Inevitably the two men would have discussed religion, and Robyns, a staunch and loyal Catholic, who in conversation probably defended Catholicism against Lollardy and the new Lutheran contagion already sweeping through Cambridge, may have been the main influence in quickening Recorde's interest in theological matters.

Robyns would have been scandalised by the activities of a certain Thomas Garrard, who was later to be counted as one of the first English Protestants. Recorde, the possessor of an open and receptive mind, may have been less shocked than Robyns when it was discovered that Garrard was the centre of a book-smuggling enterprise. He had been sending cartloads of Luther's prohibited books from the London booksellers to the universities and Recorde, with a young man's eagerness to acquire that which was forbidden, may well have been one of his buyers. Garrard was also busy distributing small, easy-to-hide copies of William Tyndale's illicit New Testament, and perhaps Recorde owned one of these too. Garrard's activities ended when he was betrayed to the authorities at Oxford. He was seized and locked in his chamber, but he picked the lock and fled.[8] By the time of this incident we can surmise that Recorde was already inculcated with the tenets of Protestantism, finding, like so many others, solace in the belief of justification by faith alone. As the veil of Roman Catholicism fell from his eyes and he unburdened himself of the centuries-old weight of Catholic dogma, superstitious ritual and unquestioning obedience, Recorde began to find himself increasingly ostracised by the other fellows of All Souls. To his dismay the society there was, as R. T. Gunther put it, proving to be 'somewhat uncongenial'.[9]

During the early years of the 1530s the intellectual world of the universities was in turmoil over Henry VIII's intended divorce from Catherine of Aragon. This raised especially difficult issues for the fellows of All Souls, many of them clergy. If the divorce succeeded it would ensure a break with Rome, end papal jurisdiction and undermine the whole of traditional theology and canon law, the two subjects of study that All Souls was founded to uphold. A royal injunction by the king in 1535 did not help. This ordered, among other resolutions, that henceforth no more lectures were to be given on canon law.

Recorde, caught up in the spirit of the day, was probably advocating the cause of the king and actively championing the Reformation. It would have been impossible for him to do one and not the other, because the two were inextricably intertwined. This could hardly have endeared him to his colleagues, whose dependence on the Roman Church was as much a matter of a future career as of faith. Should the king gain supremacy over an Anglican Church, not only would there be no need of canon law, but the development of a new theology dictated by the king rather than the pope would surely arise. The fellows of All Souls were educated and trained especially to function and gain their living by administering the traditional laws and practices of the Catholic Church, and anyone who supported change would find himself amidst very uncongenial company indeed.

Recorde must have pondered and weighed up his options. He had developed a passion for the sciences, especially mathematics and astronomy, and he now had a burning desire to study these subjects beyond the narrow limits of the quadrivium offered by Oxford. With an eye to a future livelihood outside the ecclesiastical careers now probably closed to him at All Souls, medicine must have seemed a good choice. Cambridge already had an excellent reputation for the study of medicine, and he determined to remove himself to that university forthwith, there also to pursue learning in the mathematical sciences.

There was a further compelling reason for him to leave Oxford; it was all too easy for his enemies to hurl at him the accusation of heresy. This could swiftly place him in great danger, as the case of Thomas Bilney, an Oxford professor of civil law, showed all too clearly. Bilney was brought before the bishop of London and urged to renounce his adherence to Protestantism and the opinions of Luther or be burnt at the stake. He doggedly refused to recant and was duly burnt, reputedly standing unmoved in the flames, crying out 'Jesus, I believe', the last words he was heard to utter. Recorde is likely to have perceived Cambridge as not only a good place to continue his studies, but also as somewhere to lie low for a while until the storm at Oxford had blown over.

He probably left around 1537, and he may have been heartened on his departure by knowing that the first Act of Union between England and Wales had recently come into force, opening up to the Welsh:

> all and singular freedoms, liberties, rights, privileges and laws within this realm and other the King's dominions as other the King's subjects naturally born within the same have, enjoy and inherit.

However, since these 'privileges' included the privilege of being burnt alive, it was clearly time to go.

3
CAMBRIDGE SAVANT

Recorde arrived at Cambridge as an established academic, a fellow of All Souls with a reputation for profound scholarly knowledge. There is an intriguing possibility that he took up residence in St John's College, expressly founded for the study of the liberal arts and theology and hinted at because he seems to have associated with members of that distinguished community. In any event, for the next five years or so he seems to have deliberately refrained from drawing attention to himself. He became a remarkably versatile teacher, esteemed by students for the clarity of his expositions on many difficult subjects. He indulged his passion for mathematics, probably attending whatever lectures were available and privately studying any mathematical and scientific texts he could find. In the process he educated himself far beyond the learning of most other scholars at Cambridge, saving perhaps a handful of other savants in the mathematical sciences. One such was John Dee, fifteen years his junior, with whom he became acquainted through mutual interests in mathematics and medicine.[1] Dee, who graduated BA from St John's College in 1546, achieved recognition in the reign of Elizabeth I as a highly skilled mathematician, 'clever beyond human interpretation'.

It was not unusual to study mathematics and medicine together at this time, the two sciences often being united. With an eye to a future that might bring him a regular income, which mathematics certainly would not, Recorde settled down to the demanding challenges of studying advanced medicine. He was determined to qualify at a much higher level than that achieved at Oxford, aiming for a master's degree that

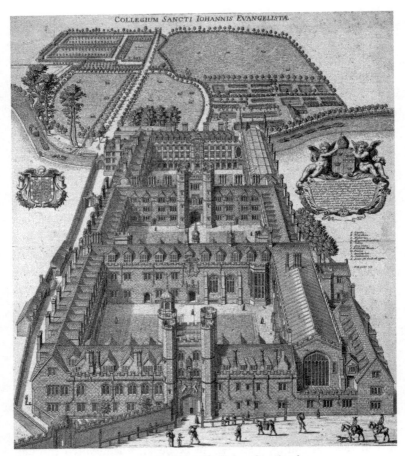

FIGURE 3   St John's College, Cambridge

Recorde possibly resided in the college, and his contemporaries Richard Whalley and Dr John Dee were both alumni. He may have lectured Whalley's eldest sons Thomas and Hugh during his residence. This engraving by David Loggan, from his *Cantabrigia Illustrato*, depicts the college *circa* 1685.

would not only provide evidence of his medical status, but also increase his capability of earning substantial fees. Presumably he continued to practise medicine as far as his Oxford licence permitted, in order to add practical experience to his academic training.

When he began his medical studies at Cambridge, Recorde would already have been well acquainted with the principle which dominated all medieval western medicine right up until the nineteenth century; the theory of humours. This asserted that within every individual there existed four fluids, called humours, which needed to be present in equal quantities, that is, they had to be in balance for a person to remain healthy. The humours were blood, phlegm, black bile and yellow bile, the theory stating that these were produced by the various internal organs in the body; blood by the liver, phlegm by the lungs and brain, black bile by the gall bladder and yellow bile by the spleen. All diseases, from simple runny noses to the most feared maladies like ague, plague and the sweating sickness, were thought to result from an excess or deficit of one or more of the humours. Too much phlegm in the body, for instance, caused breathing difficulties and diseases of the lungs. An excess of black bile caused bladder and urinary problems, not enough yellow bile engendered debilitating weakness, a surplus of blood caused fevers.

The correct balance of humours could be achieved by diet and medicines, and Recorde would have diligently studied these vital aspects of his profession. It was thought that warm foods, for example, tended to produce yellow bile, while cold foods tended to produce phlegm. Seasons of the year, periods of life, geographic regions and occupations also influenced the nature of an individual's humorous make-up. Recorde would have learnt that symptoms which seemed to be due to too much blood could be relieved by bloodletting, the opening of a vein and allowing blood to drip slowly into a bowl while the patient's pulse was assiduously measured. He would write at a later time about the possible dangers in this procedure, and warn that:

> nothing is more necessary than the exact knowledge in anatomy, to the intent you may perceive the whole course of the veins, with the like distribution of the arteries jointly passing with them, and namely in such places where blood is used to be drawn, that an artery be not stricken instead of a vein, and so danger succeed in the stead of remedy.[2]

Recorde would have attended lectures about associating the four humours with the four seasons, excess or deficiency of blood being more likely in spring, yellow bile in summer, black bile in autumn and phlegm in winter. He would have studied how these theories were closely allied to the scientific theory of the four physical elements, earth, air, fire and water, with earth predominantly present in the black bile, air in the blood, fire in the yellow bile and water in the phlegm. Thus not only physical health was influenced by the humours but also a patient's moods, and so ultimately personality, which could be affected either positively or negatively by the quality of the various fluids. People with an excess of blood supposedly had a sanguine temperament, that is, they were social, extroverted and usually courageous, hopeful or amorous. Choleric people had an excess of yellow bile which gave them energy, passion and charisma, but also tended to make them easily angered and bad tempered. People who were melancholic had too much black bile, which made them despondent, sleepless and irritable, but they could also be creative, kind and considerate. Those with a phlegmatic temperament, due to excessive amounts of phlegm, were usually calm and unemotional and characterised by dependability, kindness and affection. Recorde would have been expected to mark these bodily dispositions particularly carefully, since they determined a patient's susceptibility to particular diseases as well as behavioural and emotional propensities.

Another difficult skill all neophyte physicians had to master was the taking of pulses at all places in the body where an artery could be compressed against a bone. Using fingertips sensitised by deliberate training, pulses could be detected in the neck, on the inside of the elbow, at the wrist, in the groin, behind the knee, near the ankle and on the foot. Doctors were trained to give a prognosis for an ailment according to whether the pulse was full, easily palpable, barely palpable or bounding and irregular. It took a considerable amount of practice to be able to detect pulses and decipher their meanings, an arcane and difficult art which physicians took care to emphasise and exploit in their dealings with patients.

Uroscopy, the examination of urine as a means of assessing health and the diagnosis of disease, would also have comprised a major part

> ## OF VRINE. 18.
>
> Aboue these iij. regions, about the very brinke of the vrin you may see a certen rynge as it were going about, & that is called the crowne. Highest of al thynges in the vrine are the bubbels, which other go about with the ryng only, or els fleete in the myddell of the vrin only, or els bothe. Yea, sumtyme they couer all the whole topp of the vryne. Besyde these is there often tymes as it were a flotes, or fattynes on the topp, and sumtymes certayne spottes onely which are lyke to droppes of oyle. And these commenly ar the whole contentes. For as for grauell or stone, or any lyke thyng is conteyned vnder the name of
>
> D.ij.   Dyfforme
>
> *Crowne.*
>
> *Bubles.*
>
> *Fattynesse.*

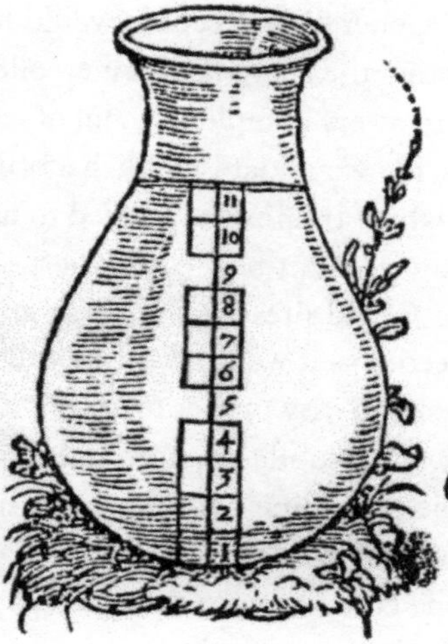

FIGURE 4 A urinal, the glass vessel used for the examination of urine
Here Recorde explains that urine (spelt vrine – the letters 'u' and 'v' were often used interchangeably in early printed books), might show 'fattiness' or 'sometimes certain spots only which are like to drops of oil'. (*The Urinal of Physick* (1547), p. 18.)

of Recorde's training. Urine, as waste matter from drink and food eaten, was thought to issue somehow from all organs and not just the kidneys, and its condition was therefore a vital indication of whole body health or degeneration. Recorde would have had to spend many hours contemplating the analysis of urine, collected in a globular urine flask, according to a very complex system of colours, layers, sediments and suspensions. He would later publish a brilliant exposition on this difficult art, but whilst training he needed to understand that the flask should not on any account be shaken. Left undisturbed in the glass vessel, the urine would present the same appearance as regards its diagnostic properties as it would if it was still in the patient's bladder. It would have been no easy matter to absorb all this medical teaching, consisting of far more subtle nuances and esoteric knowledge than it has been possible to outline here. Recorde, like all his fellow physicians, would have needed every one of the many years allotted for his training before he could be considered adept and a learned master of his profession.

Despite his demanding medical studies, Recorde was not neglecting his pursuits in the sciences. He was making inroads into astronomy, geography, the study of minerals, zoology and, it seems, just about anything else that caught his fancy. He was also deeply interested in theology, but above all he immersed himself in mathematics and his reputation for scholarship grew until, lauded by all who knew him, his prestige and celebrity seemed to have no bounds. Recorde was probably being remunerated for lecturing at this time; indeed, as lecturing was the custom, the university authorities probably expected it of him.

In moments of quiet contemplation, Recorde must often have mused on the purpose of mathematics and on the methods of teaching it. What inspired him to write a textbook on arithmetic is uncertain; perhaps it was remembrance of his father and brother back in Tenby. They were still keeping their merchant accounts in cumbersome Roman numerals and probably struggling to perform simple arithmetical calculations in the latest fashion, using Arabic numerals written with pen on paper. He determined his book would comprise three parts, the first and longer part to be 'a dialogue between the master and the scholar, teaching the art and use of arithmetic with the pen'. He saw that it was likely to be some time yet before this new method of carrying out calculations supplanted older methods, so the second part would be 'a dialogue of accounting by counters'. His third part, remembering the bargaining he had seen as a child on Tenby's harbour wall, would be a brief description of 'the art of numbering by the hand'. Recorde considered arithmetic to be the basis, the very ground, not only for the study of higher levels of mathematics like geometry and algebra, but also of every other art and science worthy of investigation. Accordingly, he titled his book *The Grounde of Artes*, (hereafter referred to as the *Grounde*) and, like all authors of the period, he knew that it was vital to its success to dedicate it to some influential person. A problem for Recorde, secluded within the sheltered walls of academe, was that he had met few people well known and exalted enough to be worthy dedicatees. It was natural therefore, that he should turn to the only one of his acquaintances with the necessary distinction; Richard Whalley.

Recorde was in his early thirties, content in the quiet and peaceful milieu of Cambridge, when he first became acquainted with Whalley, a man he came to admire and respect.[3] Whalley had risen to prominence at court during the great religious and secular changes brought about by the divorce of Henry VIII from Catherine of Aragon, his subsequent marriage to Anne Boleyn, the separation from the papacy in Rome and the establishment of an Anglican church. Whalley was a man of the world, a government servant working on behalf of Thomas Cromwell in the dispersion of former monastic lands, someone the like of whom Recorde had probably never encountered whilst cloistered in Oxford. It is no wonder that he was entranced by Whalley, and Whalley's committed Protestantism could only have endeared him the more to the young academic.

Whalley had married Laura Brookman and by her he had five children. She died about the time that he and Recorde met, but Whalley married again; Ursula, whose surname is unrecorded, gave him a further thirteen children. It has been often stated, as a matter of tradition rather than established fact, for there is no evidence, that Recorde tutored Whalley's children. It is hardly likely that he would have taught the youngest of the brood; he was after all a university lecturer, not a kindergarten teacher. It is possible though that Whalley's eldest son and heir Thomas, perhaps with his younger brother Hugh, were students at Cambridge, and if so Recorde could well have had a hand in their education. Whalley was himself an alumnus of St John's College, so it is likely that he sent his sons there, and this might explain Recorde's meeting with him, and them, in the first place.

Recorde handsomely praised Whalley for his love of learning, but in the preface to the *Grounde*, he respectfully began his address 'To the right worshipful Master Richard Whalley Esquire', by complaining about the sorry state of learning among his countrymen:

> Sore oftentimes have I lamented with myself the unfortunate condition of England, seeing so many great clerks to arise in sundry other parts of the world, and so few to appear in this our nation, whereas for excellence of natural wit (I think) few nations do match English men. But I cannot impute the cause to any other thing than to the contempt or misregard of learning. For as Englishman are inferior to no men in mother wit, so they pass all men in vain pleasures, to which they may attain without great pain or labour, and are as slack to any never so great commodity if there hang of it any painful study. Howbeit, yet all men are not of that sort, though the most part be, the more pity it is.

It is interesting to note how in this address Recorde discounts his Welsh origins and is happy to regard himself as an Englishman. Perhaps the recent emancipation of the Welsh had sunk deep into his psyche. After writing some fulsome praise of Whalley, lauding him as a man 'whose gentle nature and favourable mind is ready to receive thankfully ... this little book of numbering', and expressing the hope that 'England will (after she hath taken some sure taste of learning) ... be able to compare with any realm in the world', Recorde explains his teaching methods and the reasons for writing his book:

Therefore will I now ... return again to this my book, which I have written in the form of a dialogue, because I judge that to be the easiest way of instruction, when the scholar may ask every doubt orderly, and the master may answer to his query plainly. Howbeit, I think not the contrary, but as it is easier to blame another man's work, than to make the like, so there will be some that will find fault, because I write in dialogue. But as I conjecture, those shall be such as do not, cannot, other will not perceive the reason of right teaching, and are therefore unmeet to be further answered unto, for such men with no reason will be satisfied. And if any man objects that other books have been written of Arithmetic already so sufficiently, that I needed not now to put pen to the book, except I will condemn other men's writings, to them I answer, that as I condemn no man's diligency, so I know that no one man can satisfy every man. And therefore like as many doth esteem greatly other books, so I doubt not, but some will like this my book above any other English Arithmetic hitherto written, and namely such as shall lack instructors, for whose sake I have so plainly set forth the examples, as no book (that I have seen) hath done hitherto, which thing shall be great ease to the rude reader.

It is often stated that the *Grounde* was the first arithmetic textbook printed in England and also the first written in English. Neither statement is true, and Recorde himself was clearly aware of earlier English books on the subject. The very first arithmetic book in the country was authored by Cuthbert Tonstall, bishop of London, but this was written in Latin and so beyond the comprehension of most people. There were also at least two books already printed in English and it is these that Recorde was probably referring to in his address to Whalley. The first was printed in 1526 by Richard Fakes. It survives today as a single leaf preserved by the British Library, the verso of which contains the printer's colophon and date and also carries the words 'Thus endeth the Art and Science of Arithmetic', which may have been the title of the work.[4] The second book, entitled *An Introduction for to Learn to Reckon with the Pen, or with the Counters*, was printed by John Herford in 1536–7 (the title page gives the earlier date, the colophon the latter) at a press sited in the abbey of St Albans under the patronage of the abbot, Richard Boreman.[5] Both were translations from earlier French and Dutch texts. Although these books ran to further editions and *An Introduction* in particular sold quite well, neither could come near to claiming the phenomenal success that Recorde's book would achieve.

While writing his dedication to Whalley, Recorde could have had no conception that his treatise would be consulted by renowned statesmen, explorers, navigators, surveyors and humble craftsmen long after his death, who would be persuaded by it that mathematical knowledge was not only useful but could actually help them in their daily affairs. He would have been astonished to know that his book set the pattern for arithmetical texts that succeeded it for many years, although he expressed the hope to Whalley that some would profit by it:

> Therefore good master Whalley, though this book can be unto yourself but small aid, yet shall it be some help unto your young children, whose furtherance you desire no less than your own. And though to you privately I do it dedicate, yet I doubt not (such is your gentleness) but that you can be content that all men use it, and employ the same to their most profit. Which thing if I perceive that they thankfully do, and receive it with as good will as it was written, then will I shortly with no less kindness set forth such introductions into Geometry and Cosmography, as hitherto in English hath not been enterprised, wherewith (I dare say) all honest hearts will be pleased, and all studious wits greatly delighted. I will say no more, but let every man judge as he shall cause.

We shall see later how Recorde fulfilled his promise to write more books. He ended his dedication with words which showed his religious faith and were probably intended to emphasise his piety in the eyes of his fellow Protestant, Whalley:

> And thus for this time will I stay my pen, committing both you and all yours (good master Whalley) to that true fountain of

perfect number, which wrought the whole world by number and measure. He is trinity in unity and unity in trinity, to whom be all praise, honour and glory. Amen.

It is not difficult to imagine Recorde with tongue in cheek as he wrote the words 'all yours' in referring to Whalley's dependants. He would, of course, have been aware of his eighteen children, and probably would have been no less amused had he known that Whalley would be married a third time, to a certain Barbara, and father another seven, making twenty-five children in all. It is certain that Recorde was a droll man, and he quite unselfconsciously inserted humour into his book on arithmetic, which made it something of a rarity and special on that account alone.

4

SUCH IS YOUR AUTHORITY

The imaginary dialogues between the master and the scholar in the *Grounde* display such sympathetic understanding of the difficulties likely to beset a beginner in arithmetic as to suggest that Recorde must have had a real person in mind as an archetype for his clever and rather precocious pupil. Tradition, unsubstantiated like so much else associated with Recorde, has it that he modelled the scholar on one of Richard Whalley's children. If true, the nearest we are ever likely to come to putting a name to the scholar is that of either Thomas or Hugh Whalley. However, we should not lose sight of the fact that the scholar, the person Recorde is aspiring to teach, is actually you, the reader, and into the scholar's mouth he puts all the questions that a reader might reasonably think to ask. The master is, of course, Recorde himself. Accordingly, the dialogue of both master and scholar is actually his voice, reaching across to us over the centuries through his written texts with the clarity of a modern recording.

The *Grounde* begins with the scholar, unconvinced of the necessity of arithmetic, grumbling to the master that such simple things as numbers are a subject fit only for children.

> SCHOLAR: Sir, such is your authority in mine estimation, that I am content to consent to your saying and to receive it as truth, though I see none other reason that does lead me thereunto. Whereas else in my own conceit it appears but vain to bestow any time privately in learning of that thing, that every child may and does learn at all

times and hours when he does anything himself alone, and much more when he talks or reasons with others.

The master swiftly corrects his misunderstanding and stresses the importance of numbers.

> MASTER: Lo, this is the fashion and chance of all them, that seek to defend their blind ignorance. That when they think they have made strong reason for themselves, then have they provided the quite contrary. For if numbering be so common as you grant it to be, that no man can do anything alone, and much less talk or bargain with others, but he shall still have to do with number. This proves not number to be contemptible and vile, but rather right excellent and of high reputation since it is the ground of all men's affairs.

In order to emphasise the importance of numbers, the master poses a series of searching questions, which cause the scholar to pause and then amend his previous opinions.

> MASTER: Wherefore in all great works are clerks so much desired? Wherefore are auditors so richly fee'd? What causes geometricians to be so highly enhanced? Why are astronomers so greatly advanced? Because that by number such things they do find, which else should far excel man's mind.
>
> SCHOLAR: Merely sir, if it be so, that these men by numbering their cunning do attain, at whose great works most men do wonder, then I see well I was much deceived and numbering is a more cunning thing than I took it to be.

The master emphasises his point, determined that his pupil shall not begin his study of arithmetic without a strong sense of its usefulness and of his own ignorance of the subject.

> MASTER: If number were so vile a thing as you did esteem it, then need it not to be used so much in men's communication.

Exclude number and answer me this question. How many years old are you?

The scholar has no idea of his age and can only mumble in answer 'Mmm...'

MASTER: How many days in a week? How many weeks in a year? What lands has your father? How many men does he keep? How long is it since you came from him to me?

The scholar, never having been taught to count and tally the quantities involved in these questions, ponders and again answers 'Mmm...' The master drives home his point.

MASTER: So that if number wants, you answer all by Mmm... How many miles to London?

Defeated by this question as by all the others and aggravated by his own ignorance, the scholar replies in exasperated pique.

SCHOLAR: A poke full of plums.

MASTER: Why thus may you see what rule number bears, and that if number be lacking it makes men dumb, so that to most questions they must answer Mmm...

SCHOLAR: This is the cause sir, that I judged it so vile, because it is so common in talking every while. For plenty is no denty, as the common saying is.

MASTER: No, nor store is no fore, perceive you this? The more common that a thing is, being needfully required, the better is the thing and the more desired.

The master takes the next few pages to extol the virtues of arithmetic to the still doubting scholar. He patiently explains its essential part in the sciences, geometry and astronomy, in the practice of government, law, philosophy, the supplying and ordering of armies in times of war, and for auditors, treasurers, receivers, stewards, bailiffs and suchlike. He concludes by adding:

> MASTER: If I should (I say) particularly repeat all such commodities of this noble science of Arithmetic, it were enough to make a very great book.
>
> SCHOLAR: No, no, sir, you shall not need. For I doubt not, for this that you have said, were enough to persuade any man to think this art to be right excellent and good and so necessary for man, that (as I think now) so much as a man lacks of it, so much he lacks of his sense and wits.

The master is amused by his scholar's sudden change of perspective.

> MASTER: What, are you so far changed sides, by hearing the few commodities in general? By likelihood you would be so far changed if you knew all the commodities in particular.
>
> SCHOLAR: I beseech you sir, reserve those commodities that rest yet behind unto their place more convenient. And if you will be so good as to utter at this time this excellent treasure, so that I may be somewhat enriched thereby, and if ever I shall be able, I will requite your pain.
>
> MASTER: I am very glad of your request, and I will do it speedily, since to learn it you be so ready.

The next exchange illustrates Recorde's fondness for rhyme. His abilities as a poet will be discussed later, but here we have a first example of the rhyming prose that makes his books, despite their serious intent, so delightful to read. The scholar continues:

> SCHOLAR: And I to your authority my wits do subdue, whatsoever you say, I take it for true.

> MASTER: That is too much and meet for no man, to be believed in all things, without showing of reason. Though I might of my scholar some credence require, yet except I show reason I do not it desire.

Here is a lesson that Recorde hammers home again and again throughout this book, and throughout all his other works. Namely, the scholar should not believe everything he is told, no matter on what authority, unless it is proved to him beyond all reasonable doubt. The master then exhorts the scholar to pursue his studies diligently, and receives his assurance that he will do so.

> MASTER: But now, since you are so earnestly set this art to attain, best it is to omit no time lest some other passion cool this great heat, and then you leave off before you see the end.
>
> SCHOLAR: Though many there be so inconstant of mind, that flitter and turn with every wind, which often begin and never come to the end, I am none of their sort, as I trust you partly know. For by my goodwill what I once begin, till I have it fully ended, I never blynne.[1]
>
> MASTER: So have I found you hitherto indeed, and I trust you will increase rather than go back, for better it were never to assay, than to shrink and flee in the middle way, but (I trust) you will not so do.

The lessons begin proper with the scholar being shown how to set his numbers down in an orderly fashion, with units on the right, tens next left, then hundreds, then thousands, and so on. This may seem elementary, something which infants are taught in primary school today, but it is important to realise that the *Grounde* is not a textbook for children. For adults who had never before encountered calculations done with a pen on paper, rather than on an abacus or counting board, if at all, it was absolutely essential to begin with the simplest premises and Recorde well understood this.

MASTER: So that arithmetic is a science or art teaching the manner and use of numbering, and may be wrought diversely with pen or counters, and other ways. But I will first show the working with the pen, and then the others in order.

SCHOLAR: This I will remember. But how many things are to be learnt, to attain this art fully?

The master answers by proceeding with lessons in the arithmetical arts of addition, subtraction, multiplication, division, reduction and progression. Each process is explained in great detail, with interpolations by the scholar, who frequently appears not to understand, to misunderstand, or to get things just plain wrong. This, of course, is a teaching strategy skilfully employed by Recorde, allowing the master to amplify his discourse and explain difficult topics in more than one way.

As an example of how Recorde teaches mathematics, we might consider his method of multiplying simple numbers. Most people at the time could manage numbers under five, but multiplying digits greater than that could be a problem. It must be realised that 'times tables' were not committed to memory as is the practice with young children today, so rules for single digit multiplication had to be mastered before learning how to multiply greater sums.

SCHOLAR: Therefore sir, I beseech you, teach me the working of it.

MASTER: So I judge it best, but because that great sums cannot be multiplied, but by the multiplication of digits, therefore I think best to show you first the art of multiplying them, as when I say, 8 times 8, or 8 times 9, etc. And as for the small digits under 5, it were but folly to teach any rule, seeing they are so easy that every child can do it. But for the multiplication of the greater digits thus shall you do. First set your digits one over the other right, then look how many each of them lacketh of 10, and write that against each of them, and that is called the differences. As if I would know how

Multiplication.

1	1	2	3	4	5	6	7	8	9
	2	4	6	8	10	12	14	16	18
		3	9	12	15	18	21	24	27
			4	16	20	24	28	32	36
				5	25	30	35	40	45
					6	36	42	48	54
						7	49	56	63
							8	64	72
								9	81

In whiche figure when you wolde knowe any multiplicatiō of digettes, seke your fyrste or laste digette, in the blacke squares, and from it go ryght forth towarde the ryghte hande, tyll you come vnder the figure of your seconde diget whiche is in the hyghest rowe, and then the nomber that is in the metynge of theyr bothe squares, is the multiplycation that amoūteth of them. As yf you wolde knowe by this table the multiplicatiō of 7 tymes 9, seke fyrst 7 in the blacke squares, and then go ryghte forth toward the

FIGURE 5 Recorde's table for multiplying digits

When you would know any multiplication of digits, seek your first or last digit in the black squares, and from it go right forth toward the right hand, till you come under the figure of your second digit which is in the highest row, and then the number that is in the meeting of their both squares, is the multiplication that amounteth of them. (*The Grounde of Artes* (1543), p. 49v.)

many are 7 times 8, I must write those digits thus.

$$\begin{array}{c} 8 \\ 7 \end{array}$$

Then do I look how much 8 doth differ from 10, and I find it to be 2, that 2 do I write at the right hand of 8, thus.

$$\begin{array}{cc} 8 & 2 \\ 7 & \end{array}$$

Then do I take the difference of 7 likewise from 10, that is 3, and I write that at the right side of 7, as you see in this example. Then do I draw a line under them, as in addition, thus.

$$\begin{array}{cc} 8 & 2 \\ \underline{7} & \underline{3} \end{array}$$

Then do I multiply the two differences, saying 2 times 3 makes 6, that must I ever set under the differences beneath the line. Then must I take that one of the differences (whichever I will for all is like) from the other digit, not from his own, and that that is left, must I write under the digits, as in this example.

$$\begin{array}{cc} 8 & 2 \\ \underline{7} & \underline{3} \\ 5 & 6 \end{array}$$

If I take 2 from 7, or 3 from 8, there remains 5, that 5 must I write under the digits, and then there appears the multiplication of 7 times 8 to be 56. And so like of any other digits, if they be above 5, for if they be under 5, then will their differences be greater than themselves, so that they cannot be taken out of them, and again such little sums every child can multiply, as to say, 2 times 3, or 4 times 5, and such like.

SCHOLAR: Truth it is, and seeing, me seemeth, that I understand the multiplying of the greater digits, I will prove by an example how I can do it.

This the scholar does to the master's satisfaction, after which he is provided with a 'table to multiply all digits by, for your ease and surety in working'. The exchange of dialogue is adroitly and delightfully handled, by which means we see the scholar's initial ignorance slowly yielding to enlightenment, and his doubts giving way to growing confidence. By illustrating the scholar's burgeoning competence in this way, Recorde artfully places in the minds of readers the notion that they too could succeed in this seemingly difficult art of arithmetic. This is surely the sign of a great teacher. Practice exercises follow, an instance being when the master, after having demonstrated how to add simple columns of numbers, broaches the more difficult problem of adding together sums of money. He shows the scholar how to do it and then invites him to do likewise.

MASTER: And this may you prove in another like sum.

SCHOLAR: Then will I cast the whole charge of one month's commons at Oxford, with battelling also.

MASTER: Go to, let me see how you can do.

One can imagine Recorde, sitting in his chamber at Cambridge as he composed his great arithmetical text, pausing here and reflecting on his days at Oxford. Drawing on his remembrance of Oxford's unique terminology, commons and battelling, for the system of communal and personal bookkeeping, he works this up into an exercise for the scholar. At the beginning of the *Grounde* he had already explained the 'figures of money', comprising pounds, shillings, pence, half-pence, quarter-pence (a farthing), and the further subdivision into eighth-pence (a kew) and sixteenth-pence (a cee), the latter denominations peculiar to Oxford. These species of money are represented by Recorde, without resorting to fractions, with the respective symbols li, s, d, ob, ö, q and c. The scholar's reckoning is thus:

> SCHOLAR: One week's commons was 11d ob ö and my battelling that week was 2d ö q. The second week's commons was 12d and my battelling 3d. The third week's commons 10d ob and my battelling 2d q c. The fourth week's commons 11d ö and my battelling 1d ob c. These four sums would I add into one whole sum, and therefore I will set them one over another thus.

The scholar then lists the weekly amounts, one under the other, carefully keeping each denomination in its own column. He adds the columns, carrying forward to the next column any whole sums, as he has been taught, and finally produces his answer.

> SCHOLAR: Then appears all my addition thus. And the sum is 4s 6d ob q [4s. 6⅝d].
> MASTER: Now have you done this well.

Methods by which the correctness of long and difficult additions, subtractions, multiplications and divisions can be proved are explained to the scholar in the most painstaking way. At one point, the scholar becomes confused about the method of proving the addition of money.

> SCHOLAR: I perceive [the method] does serve for those three denominations, pounds, shillings, pennies. But what if I had halfpennies, farthings, kews and cees?

The master is amused, because kews and cees are not something the scholar will encounter in everyday life.

> MASTER: You think you be at Oxford still, you bring forth so fast your kew and cee.

The master gives the scholar the necessary guidance and the dialogue continues, with ample explanations, demonstrations and exercises at every stage, through the arts of subtraction, multiplication, and division.

By the later stages of the book the scholar has shown himself to be a fast learner, at times even brilliant, by his rapid comprehension and ready ability to apply what he has learnt. This is a deliberate pedagogic tactic by Recorde, designed as before to persuade his readers that if the scholar can do it, so can they. The master is now ready to test the scholar's cognition with more difficult and searching questions.

> MASTER: If I sold unto you a horse, having four shoes, and in every shoe six nails, with this condition, that you shall pay for the first nail one ob, for the second two ob, for the third four, and so forth doubling unto the end of all the nails. Now I ask you, how much would the whole price of the horse come unto.

The scholar, nothing daunted, leaps in to give his answer.

> SCHOLAR: First to know the number of the nails, I must multiply 6 by 4, and that makes 24, then I will do thus. I will write the number of the nails every one in order from one to twenty-four, and against each number of nails the sum of half pennies due.

The scholar does this, and by applying to his column of numbers a technique he has previously been taught, the rule of progression, he arrives at the surprising sum of 34,952 li 10s 7d ob (£34,952 10s 7½d). Recorde again shows his wry sense of humour in the exchange which follows.

> MASTER: That is well done, but I think you will buy no horse of the price.
>
> SCHOLAR: No sir, if I be wise.
>
> MASTER: Well then, answer me to this question. A lord delivered to a bricklayer a certain number of loads of bricks, whereof he willed him to make twelve walls of such sort that the first wall should receive two thirds of the whole number, and the second two thirds of that, that was left, and so every other two thirds of that that remained, and so did the bricklayer. And when the

twelve walls were made, there remained one load of bricks. Now I ask you, how many loads went to every wall, and how many loads was in the whole?

The scholar is completely nonplussed by the difficulty of this question, and this time not so ready to leap to an answer.

> SCHOLAR: Why sir, it is impossible for me to tell.
> MASTER: Nay, it is very easy if you mark it well.

When the master shows him how to obtain the answers sought for, by the application of logical reasoning and the mathematical knowledge he already has, the scholar is both amazed and elated.

> SCHOLAR: Lo, now it appears easy enough. Now surely I see that arithmetic is a right excellent art.
> MASTER: You will say so when you know more of the use of it, for this is nothing in comparison to other points that may be wrought by it.
> SCHOLAR: Then I beseech you sir, cease not to instruct me further in this wonderful cunning.

Recorde ends the first and major part of the *Grounde* and then begins the second dialogue, of accounting by counters, by having the master address the scholar with these words:

> MASTER: Now that you have learned the common kinds of arithmetic with the pen, you shall see the same art in counters. Which feat does not only serve for them that cannot write and read, but also for them that can do both, but have not at some times their pen or tables ready with them.

There follows a fifty-page discourse on how to manipulate counters, which resemble draughts pieces on a board ruled with lines representing units, tens, hundreds, thousands, tens of thousands and hundreds of thousands. The scholar, as we have come to expect, quickly absorbs each lesson, asking questions at salient points and receiving further elucidation that is much to the advantage of the reader. The master closes the second dialogue, but Recorde is not yet finished and throws in a final topic before ending the book.

> MASTER: But one feat I shall teach you, which not only for strangeness and secretiveness is much pleasant, but also for the good commodity of it right worthy to be well marked. This feat hath been used above two thousand years at the least, and yet was it never commonly known, especially in English it was never taught yet. This is the art of numbering on the hand, with divers gestures of the fingers, expressing any sum conceived in the mind.

Recorde was recalling here what he had probably seen on the harbour wall during his childhood in Tenby, sailors and merchants speaking different languages but negotiating prices and tariffs by means of signing on their fingers. And so, after a six-page explanation of how to number with the hand, the master and scholar engage in a final exchange.

> MASTER: And now for this time farewell, and look that you cease not to practise that you have learned.
>
> SCHOLAR: Sir, with most hearty mind I thank you, both for your good learning, and also your good counsel, which (God willing) I trust to follow.

As Recorde laid down his pen he must have sighed with relief at completing such a monumental work of over 300 pages. The mental effort required of him to do so is extraordinary, considering that he would have been studying for his medical degree at the same time. However, his immediate thoughts must have turned to getting his book printed, and for that he needed to visit London, the home of the book printers.

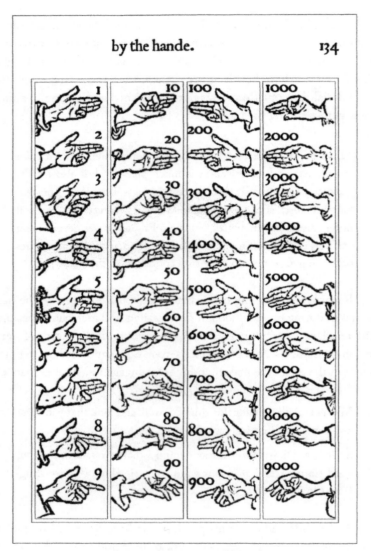

FIGURE 6 Numbering by the hand

Recorde probably saw this method of signing exchanged between merchants and seamen during his childhood in Tenby. He promised to teach 'addition, subtraction, multiplication and division' by finger movements 'which yet were never taught by any man as far as I do know.' Perhaps thankfully, he failed to keep his promise. (*The Grounde of Artes* (1543), p. 134.)

5

ST PAUL'S CHURCHYARD

Old St Paul's, the magnificent medieval cathedral that stood on the site of the present St Paul's before it was destroyed by the great fire of London in 1666, had nestling in its shadow a number of houses and small shops collectively known as St Paul's Churchyard. This was not only the hub of the London book trade but also the centre for the whole country, there being few printing presses at that time outside the city. The houses in the yards and alleys belonged to the printers, who set up their presses on the ground floor and usually lived on the floor above. Most of the shops housed booksellers and bookbinders, and many printers also incorporated a shop in the front room or basement of their premises. St Paul's Churchyard was a place of bustling activity, with wagoners and cartmen constantly delivering bales of paper, glazed pots of ink and many other essential requirements for the non-stop printing presses.

The north and south transepts of the cathedral, colloquially known as Paul's Walk, were used by those just passing through as a general thoroughfare, as well as by those congregating in search of gossip or to buy books. The cathedral was the centre of the London grapevine, and newsmongers gathered there to pass on the latest news about current affairs, war, religion, parliament and the royal court. Infested with beggars and thieves, Paul's Walk was also a place to pick up current rumours, topical jokes and, it was said, the pox. Huge crowds periodically gathered at St Paul's Cross, an outdoor covered pulpit from which the king's proclamations were made and from which the leading prelates of the time expounded, often controversially, on theology and politics.

Figure 7 The Norman Cathedral of St Paul's
Upon the Dissolution of the Monasteries, many former Catholic buildings in the churchyard, such as chapels, crypts, shrines and charnels, were seized by the Crown and sold as shops and rental properties to printers and booksellers. Reyner Wolfe had his printing house here, under the sign of the Brazen Serpent.

The throng in the streets was good for business, but overcrowding had obliged many printers and booksellers to overspill into neighbouring Paternoster Row, Fleet Street and Cheapside.

When Recorde first set foot in this quarter of London he must have been overwhelmed by the hubbub and general tumult compared with the quiet and peaceful sanctum of his normal university milieu. He was probably already familiar with the names of many printers through his extensive reading, since all of them put their names and places of business prominently inside the books they published, either on the title page or in a colophon at the end. Every printer hung a unique sign outside his house, rather like an inn sign, which, in the absence of any orderly system of street naming and house numbering, indicated the whereabouts of his premises. On his first visit to St Paul's Churchyard, Recorde sought out one sign in particular. Whether this was by previous appointment or whether it was a speculative approach on his part we do not know, but he was eventually to present himself at the sign of the Brazen Serpent.

It was under this sign that a foreigner, a native of Gelderland in the Netherlands, carried on a printing and bookselling business. This was Reyner Wolfe, who also spelled his first name as Reynar or Reynold, or in an anglicised version as Reginald. Wolfe was an intellectual and a strongly committed Protestant who had initially settled in Strasbourg, where he learnt and practiced the art of printing. While there he established a network of European connections, both in the book trade and also among many prominent theologians and Lutheran religious reformers. One of his correspondents was Simon Grynaeus, a German scholar and theologian of the Protestant Reformation, who visited England in 1531 to carry out researches in monastic libraries. It is likely that it was Grynaeus who brought Wolfe to the attention of Thomas Cranmer, then rapidly becoming the recognised leader of the English Reformation. Cranmer was appointed Archbishop of Canterbury in 1532 and the following year he was instrumental in bringing Wolfe to England, no doubt perceiving in the Dutch Protestant a useful contact with the Lutheran and Swiss reformers on the continent.

Wolfe's first act on his arrival in London was to take out a patent of denization. The patent was an exercise of the royal prerogative by which a foreigner could obtain certain rights otherwise enjoyed only by the king's subjects, such as the right to hold land and acquire property. Wolfe paid his fee and took the oath of allegiance to the Crown, not becoming a full citizen but no longer to be regarded as an alien, his status being somewhat akin to that of a permanent resident. Wolfe immediately began to exploit his network of contacts to import the works of continental publishing centres for sale in the English market. For this purpose he travelled annually to the great book fair at Frankfurt am Main and the books he purchased there were shipped back to London wrapped in straw and packed in barrels. This may seem a strange method of packaging, but considering how heavy and difficult to lift a wooden crate filled with books would be, compared to the comparative ease with which a barrel could be laid on its side and rolled by a single man, the method was both logical and practical.

Wolfe had need of great caution over the books he imported. All Lutheran works, whether in Latin or in English, had been declared heretical in England and arrests for infringement of this embargo were not infrequent. Such a case occurred in 1531, just prior to Wolfe's arrival in England, when a certain Richard Bayfield was taken, charged with importing prohibited books, and on 14 December of that year burnt at the stake.[1]

Cranmer readily utilised Wolfe's frequent trips to the continent to further his own religious and political ends, as well as those of the king's Privy Council. In 1536, when Cranmer wished to reopen contact with the Swiss reformer Heinrich Bullinger, then head of the Zurich church and one of the most influential theologians of the Protestant Reformation, Wolfe, in conjunction with the Zurich printer Christopher Froschover, carried friendly messages back and forth between the two men. A few years later Wolfe received a payment of 100 shillings, a not inconsequential sum, for conveying letters between Henry VIII and his agent in Germany, Christopher Mounte.[2] Wolfe was now firmly established in the role of spy and was expected to keep his ear close to the ground. This was perhaps not espionage as we would later understand the term, but rather involved him as a carrier back to England of tittle-tattle garnered among people of influence which might be of interest to his paymasters. Nevertheless, this could be a dangerous business in a time of religious dissent and political turmoil when it was all too easy to cross the wrong factions, but Wolfe seems to have survived unscathed.

In 1538 Nicholas Partridge, an associate of Cranmer, wrote to Bullinger telling him that 'our friend Reyner did not come to this fair (Frankfurt) by reason as I understand of the recent death of his wife'.[3] Despite this setback, Wolfe pushed ahead with his plans to become a printer in his own right. Cranmer supported him, foreseeing Wolfe's future usefulness when it came to the printing of his own religious theses and tracts, but there was a difficulty to be overcome before Wolfe's ambition could be realised. The Worshipful Company of Stationers, one of the livery companies of the City of London, held a monopoly over the printing and publishing trades and was officially responsible for setting and enforcing rules of conduct and the regulation of business. In 1529, concerned by the number of aliens printing in London to the detriment of native-born Englishmen, the company succeeded in obtaining an Act of Parliament that henceforth prohibited any alien from setting up a press, although it did not forbid those already established from continuing to print.

As a denizen, Wolfe should have been exempt from the provisions of this Act, but the passing of a law was one thing, its unbiased application was quite another. Obstacles were placed in his path and he appealed to Cranmer for help. The archbishop in turn informed the most powerful woman in the land, Anne Boleyn, of Wolfe's predicament, and with the queen of England's intervention his difficulties were speedily resolved. In 1536, shortly before Anne's fall and execution, Wolfe was admitted a freeman of the Stationer's Company and thus was able to bring his plans to fruition. By 1542 his printing house was firmly established under the sign of the Brazen Serpent in St Paul's Churchyard.

A major problem for Wolfe, as for all printers, was what to print. At this point in his new enterprise Recorde fortuitously appeared and placed the manuscript of the *Grounde* in his hands. Wolfe was already engaged in printing some religious works, a popular mainstay of all publishers at this time but presumably, seeing merit in Recorde's mathematical textbook, he would have welcomed it as a far less dangerous text to print than a theological or devotional treatise. Wolfe would have been well aware that John Gough had been sent to the Fleet prison the previous year for printing and selling seditious and heretical books. His near neighbour Thomas Berthelet was also in trouble for printing translations of three banned works by Erasmus, so Recorde's book must have seemed a safe and very innocuous treatise that might, perhaps, also make some money.

It was almost certainly the astute Wolfe who persuaded Recorde to lengthen the title of his book, so that it became, in the spelling of the time, *The Grounde of Artes teachyng the worke and practise of Arithmetike, moch necessary for all states of men. After a more easyer & exacter sorte, then any like hath hytherto ben set forth: with dyvers newe additions, as by the table doth partly appeare*. Such verbose titles were used by printers to advertise their wares, and books of this period typically had titles incorporating phrases such as 'newly and most truly corrected', or 'much necessary to be read of all the king's subjects', or 'necessary and needful for every person to look in' and so on. Many printers, including Wolfe, became so carried away by this means of advertising that they forgot the importance of brevity in this respect, and thus often destroyed the effectiveness of the title page by flooding it with type. Of course Wolfe, ever the salesman, could not afford to ignore such an advertising opportunity, and so Recorde's book on arithmetic had to be described as 'easier' and 'exacter' than any similar book previously published. The first edition of the treatise appeared in 1543, but Wolfe's little white lie that it contained 'diverse new additions' has misled many people into postulating and searching for an earlier edition, erroneously reasoning that only something pre-existing can have 'new additions'.

Recorde must have been amazed by the technology he saw, probably for the first time, on his initial visit to Wolfe's printing house. The press itself was a wooden structure, about six feet in length and three feet wide, with two massive oak supporting posts standing about seven feet tall.[4] We do not know how many presses Wolfe owned, perhaps only one but probably more, each capable of producing over 3,000 impressions per workday. Type, comprising small metal letters, was arranged into pages by a compositor and placed in a wooden frame, which was itself placed onto a flat stone bed for inking. This was done using two ball pads, each stuffed with sheep's wool and mounted on handles, to which the ink was first applied before being transferred evenly to the type. A sheet of paper was then placed on the surface of the inked type and the bed was rolled under a movable platen, using a windlass mechanism. The impression was made with a screw that transmitted downward pressure to the platen through the agency of a long lever controlled by a pressman. The screw was then reversed, the windlass turned again to move the bed back to its original position and the printed sheet removed.

The ink used was a sticky adhesive substance, comprised mainly of lampblack and linseed oil, the preparation of which required some scientific knowledge. It was slow drying, and whenever Recorde visited the Brazen Serpent he could hardly have failed to notice the multitudes of freshly printed pages hung up to dry like washing on overhead racks. The improvement of inks was a continual preoccupation of printers and was an important reason why men with scientific training like Recorde were always welcome on the premises. Even more amazing to him would have been the cases, shallow wooden trays stacked one above the other and divided into compartments to hold the metal type. A printer would necessarily own tens of thousands of letters in many different fonts and point sizes, each set called a fount and having the capital letters stored in the upper case and the small letters with numbers and punctuation characters stored in the lower case.

As the composition of Recorde's manuscript went ahead, galley pages of proof would from time to time be printed and any necessary corrections made. Early printing practice recognised the right of the author to oversee the printing of his work, but in any case Wolfe would have been glad to have a competent scholar like Recorde attending the press to scan the proofs as they appeared. We might suspect that Recorde was not as diligent as Wolfe probably hoped, but we need to remember that he was still heavily involved at Cambridge, completing his medical degree. The first edition of the *Grounde* contained a page of errata, that is errors that were not noticed until it was too late and all the leaves had been printed. The only recourse then was to print the errata and slip the page in at the end before all the leaves were collated and bound. Nevertheless, a few errors by the compositors, such as placing figures in the wrong column or on the wrong line, still escaped detection and must have confused potential learners when the book finally reached their hands. However, it should be borne in mind that most books at this time contained lists of 'faultes escaped', which usually filled up the last leaves and for which most authors were quick to blame the 'naughty printer'.

A normal print run for a textbook like the *Grounde* would have been between 300 and 500 copies and each copy might have been sold for between five and ten shillings. It is difficult to ascertain an accurate selling price for books at this time, as it was quite normal for printers to sell volumes bound either between wooden or pasteboard covers or as loose quires that the customer could take to a bookbinder to be enclosed within material of their own choosing, such as leather or parchment. Obviously, the prices in the two cases would vary considerably. A textbook like Recorde's, deemed to be of little intrinsic value and therefore disposable, might even have been sold without any covers at all. People unable to come to London to buy a particular book, or without an acquaintance in the metropolis to obtain it for them, were obliged to rely on the provincial booksellers. These men ordered from the London printers such books as they thought they could sell, but would of course be willing to obtain any title to fulfil a specific customer requirement.

By the time Wolfe completed printing the *Grounde* a religious campaign was in full swing against the printing, importing or selling of any books that attacked the church or disseminated the doctrines of Luther and his followers. As early as 1520 Pope Leo X had issued a Bull against the writings of Luther that ordered the confiscation and burning not only of his texts, but also of anything else found on examination to be heretical. Already on two separate occasions, with great ceremony and much preaching, books declared heretical had been publically burnt at St Paul's Cross in full view of the London printers. The message would not have been lost on any of them, and Wolfe in particular, known for his strong Protestant convictions, would have been careful to avoid any suspicion of clandestine printing or importing of forbidden works. In this he was evidently successful and he remained in favour with the king, so much so that later, in January 1547, he was appointed King's Printer in Latin, Greek and Hebrew, with an annuity of twenty-six shillings and eight pence. This could hardly have been a valuable monopoly, since the volume of printing in these three languages was very small, but it showed that Wolfe was a trusted member of the printing and bookselling fraternity at this time.

In 1538 a proclamation was read from the outdoor pulpit at St Paul's Cross declaring that *nothing* was to be printed until it had been examined and licensed by the Privy Council or its agents, such as the bishops of London, Canterbury or Winchester. Accordingly, Wolfe would have submitted Recorde's arithmetical textbook for approval and the censors would no doubt have rubber-stamped its acceptability, seeing that it dealt not with any controversial religious subjects but with mundane mathematics. If so, they missed noticing a feature inset within the title page iconography that the king himself would have regarded not only as heretical but also seditious, had it been brought to his attention. More will be said about this later, but Wolfe was surely foolhardy in allowing it to be printed. He and Recorde had become firm friends by this time, not least because of their shared Protestant faith, and he risked involving his author in possible serious repercussions. Nothing untoward did occur, but the flames of the Smithfield fires came perilously close to Wolfe and Recorde in October 'of the year of our Lord Christ 1543', when *The Grounde of Artes* finally appeared.

6

DOCTOR RECORDE

In the early part of 1545 Recorde was finally granted his MD by Cambridge University.[1] No doubt delighted by the successful conclusion of his medical studies, he might not have worried too much that his award was slightly irregular. In approving his degree the medical faculty:

> granted in the case of Robert Recorde that twelve years medical practice, after he was licensed by Oxford to begin such practice, shall entitle him to become a regular member of the medical faculty, provided he shall, before Easter, attend certain discussions, such as the Medical Meeting.

We can be sure that he did whatever was required to enable him to add the coveted designation 'Physician' after his name, which he proudly did for the rest of his life. However, it seems that he did not properly complete all the stages leading up to his doctorate, since we find the medical faculty also noting that:

> the same [Recorde] is excused from taking, but is given credit for, a certain course required for the Bachelor's degree in medicine, since, being now a Doctor, he could not with propriety take the course; and since other Doctors in the past have not been required to take it.[2]

This apparent laxity on the part of the faculty would not, in the end, benefit Recorde, as we shall see.

It has long been debated how Recorde spent the two years immediately after he qualified as a physician. One suggestion is that he spent his time teaching, and so long as he remained at Cambridge this would not be surprising, as he probably pursued the pedagogical interests in mathematics and the sciences that he had developed at Oxford. Another traditional theory is that he left Cambridge and returned to his *alma mater* in order to teach medicine. This assertion can probably be elevated from a possibility to a likely truth, since some scanty evidence exists to show that he did indeed return to Oxford for this purpose. William Bullein, a parson and a physician, claimed to have attended Oxford University at the time Recorde would have returned there. In March 1562, after Recorde's death, Bullein published a medical treatise entitled *Bullein's Bulwarke of Defence Against all Sickness, Sornes and Wounds*. In this lengthy, rambling text Bullein scatters odd scraps of information about his life and, referring to his own medical education, he mentions 'R. R. [Robert Recorde] ... under whose banner I served most and got all I have'.

Thus we glimpse Recorde attempting to establish himself as a teacher of medicine, and probably claiming his old place among the fellows of All Souls. However, after his burgeoning acquaintance with Reyner Wolfe he was more than ever a fervent and outspoken proponent of Protestantism, rendering him even less welcome among the conservative factions of the college than he was before his long sojourn at Cambridge. He must have quickly realised that there was no future for him as a university lecturer and, since he needed a profession that would secure him an income, it was natural that his thoughts should turn towards the practice of medicine for a living, rather than as an academic discipline. He therefore made the important decision to leave academia, which had cosseted him virtually from his childhood, to seek his future in London as a doctor.

Once the wrench of leaving a familiar environment was past, the future must have seemed very promising for Recorde. It has been estimated that in London at that time, out of a population of nearly 200,000, there were about five hundred medical practitioners in total, of which only about fifty were university-trained physicians, one hundred were licensed surgeons, the same number were apothecaries, and the rest were unlicensed healers and midwives. The vast majority of people in the city depended on the last group for their medical care. These people had no professional training but did have a reputation as wise healers who could diagnose problems and advise sick people on what to do, perhaps set broken bones, pull a tooth, dispense traditional herbs and brews and maybe even perform a little magic to cure their ailments. With such a scarcity of trained doctors Recorde could have anticipated a wealthy clientele, well able to afford the consultation fees that his social status and academic rank would command.

It was quite usual for physicians in London to hold consultations at conveniently located inns and taverns, but whether Recorde did so or not, it was obviously imperative that he find somewhere to live.

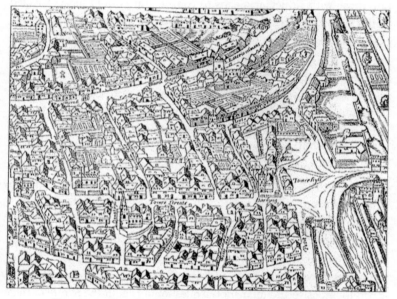

FIGURE 8 Parish of St Katherine Coleman, *circa* 1550

The church of St Katherine Coleman is the building with the tall tower just right of centre, towards the top. One of the houses adjacent to the church, many of which seem to have large rear gardens, may have been the London residence of Dr Robert Recorde. Tower Hill (spelt Towrehyll), with its permanently erected gallows, is middle right, with the Tower of London partly visible at bottom right.

He settled on a house in the parish of St Katherine Coleman, in a small maze of dwellings surrounding the church, close to the Tower of London and near Tower Hill with its permanently erected gallows.[3] He was obviously pleased with his new-found independence, proudly declaring in a medical textbook he soon set about publishing, which will be discussed below, that he was writing on 8 December 1547, 'at my house in London'. Since his was a bachelor household, Recorde would have been in need of servants, probably acquiring at this time a personal manservant whose name we only know as John.[4] It would have been usual for a person in his position to employ also a cook and a housekeeper, and probably a boy to run errands and do odd jobs.

It is interesting to consider what the citizens of London, at any rate those with sufficient means to consult him, would have expected from Doctor Recorde. Today we rely on potent and effective medicines, expecting certain drugs to be equally beneficial for all patients suffering from the same symptoms, but the principles of learned medicine were very different in Recorde's time. Physicians were not there to fight disease with pills, libations and decoctions, but rather in the first place to preserve health and prolong life by means of good advice. Unlike modern doctors, a Tudor physician would begin his diagnosis by considering the unique qualities of his patients rather than the similarity of their symptoms. The connection between health and virtue would be stressed and patients would be given advice on how to regulate their lives; thus restoration of health was expected through reformation of habits, rather than medication.

The concept of diagnosis through physical examination was hardly known and physicians would rarely touch their patients, beyond feeling the pulses of the heart, brain and liver. Instead the sick would be subjected to intensive and inquisitorial questioning about the nature of their temperament and emotional state, their constitution and strength, their infirmities, their patterns of sleep and waking and their everyday diet. Only when a doctor knew all the personal qualities and normal habits of a patient, and how these qualities and habits differed during illness, was he able to advise on the actions required to recover health. This might necessitate the administering of medicine, but usually involved pertinent advice on accommodating one's life to the needs of one's proper constitution. Preventing trouble was considered a more preferable option than a cure, and even when matters went wrong, the best remedy was reliable advice on how to return to a state of harmony and well-being.

All this considered, we might wonder what made people trust any physician to give them the correct advice they often sought so desperately. Perhaps it was the grave and dignified manner, the dark and sober dress topped for special occasions by the long gown of the university man, the Latin speech and ponderous silences when in consultation with colleagues, all of which betokened men of learning and virtue. Physicians used the word 'profession' with regard to themselves but to no other medical practitioner, thus clearly identifying their calling as something more than a mere occupation. Just like lawyers and the clergy, members of the other two professions prominent in the Middle Ages and continuing into Tudor times, the professional authority of physicians rested on the key concepts of judgement and advice. Such concepts were considered attributes of character as much as of knowledge, which alone could not result in sound judgment and advice. The purpose of the long university education that marked out physicians from all other medical men was not merely the transmission of information about healing, but the transformation of the student doctor into a physician of exemplary character, able to exercise unerring judgement and unfailing prognostication. The long and arduous education in classical languages, methods of reasoning, moral and natural philosophy, the studying of medical texts formally through disputation rather than through gaining experience by observing patients, in all of which Recorde was steeped and which had shaped his mind and his character, fitted him superbly for his life's chosen work. In the end, however, as we shall see, things did not quite work out as he had planned.

Comfortably lodged in his house in London, Recorde sat down to put the finishing touches to a book he must have started whilst still at Cambridge. It is likely that, foreseeing the need to enhance his academic reputation with a published medical treatise but much more interested in his mathematical work and producing a second and third edition of his textbook on arithmetic, he composed the contents with somewhat less than his usual rigor. His book is a traditional text on uroscopy, that is, the diagnostic examination of urine by simple inspection. In its tone it was much less modern than his mathematical text and it is not at all critical of authority. Throughout the book Recorde quotes the opinions and teachings of classical authors such as Hippocrates, Galen, Avicenna and many others without criticism or amendment. This is in marked contrast to his mathematical works, both the editions already published and books yet to come, in all of which he urges his readers to use their own reason and observation as a guide to truth before uncritically accepting the teachings of ancient authorities.

Nevertheless, the sheer range of classical authors he quotes is a testament to his own wide reading and learning in medicine. He entitled the book *The Urinal of Physick*, and it would prove, in its field, every bit as successful as his first mathematical book, remaining in print for well over a century. In the preface Recorde exemplifies the ethos of Tudor medicine when he writes:

> I have written this little treatise to all men in common, that they may learn to have some knowledge of their own urines, and thereby may be the better able to instruct the physician, in this thing at least, what sort of urine they have made from time to time, since the beginning of their sickness, and somewhat before. And also what sort of water they were wont to make customarily in their health, so that if men will be diligent to mark their water in time of health, they shall not only be able to instruct the physician (as I have said) but should be also able to perceive the cause of the disease sometimes before the grief comes, and so by counsel of some discreet physician, avoid the sickness before it be fully entered.

It is interesting to note that in a time long before political correctness, Recorde uses the word 'men' in a generic sense to mean 'men and women'.

The urinal in Recorde's title did not have quite the same meaning as is usually attached to the word today. It was instead a glass vessel in which the physician, by holding it up to the light, could inspect a patient's urine, carefully noting characteristics such as colour, thickness and sediment. Coupling these observations with age, gender and other considerations such as whether the urine was cold or still warm, he would then draw his conclusions. In his book, Recorde specified the ideal vessel for this purpose:

> Now as touching the urinal, it should be of a pure clear glass, not thick nor green in colour, without blots or spots in it, not flat in the bottom, nor too wide in the neck, but widest in the middle and narrow still toward both the ends, like the fashion commonly of an egg, or of a very bladder being measurably blown (for the urinal should represent the bladder of a man). And so shall everything be seen in his due place and colour.

Recorde went on to caution that the patient should not on any account make urine in any other sort of vessel and afterwards pour it into the approved urinal before bringing it to the physician, 'for that causes deceit and error in the judgement of it'. Furthermore, he exhorted everyone not to mock and jest with the physician by bringing to him the stale of a beast instead of the urine of a man, or the water of a man pretending it to be that of a woman. Because, he insisted, 'they thereby only deceive themselves, in so much as water from a man might declare certain health while if it were a woman's it might indicate a disease'. Contrariwise, 'water that on examination would show health if it came from a woman might indicate sickness if it came from a man'. Therefore, he wrote, 'if you genuinely seek the patient's well-being, receive the urine diligently and as soon as you can, present it to the physician, being sure at the same time to tell him all things necessary and all that he needs to know'.

The main part of the book concerns what Recorde called the 'judicial of urine', that is, how to diagnose symptoms from its colour, thickness, taste and smell. He identified six principal colours, namely white, pale, flaxen, yellow, red and black. These colours could be further subdivided, for example white comprised 'clear as crystal', 'white as snow', 'pure as water' which are the light whites, then 'milky white', 'clear like horn', and 'grey' which are dark whites. Similarly, all the other principal colours could be subdivided, each subdivision indicating some particular element of disease, sickness or health. Recorde was unequivocal about black urine, attributing its colour to morbid blood that undoubtedly 'betokened death'.

He went on to explain that urine was not only the basis from which prognoses could be made, but that it had medicinal and curative properties too. If, he wrote, 'a man let his own urine drop upon his feet in the morning, that was good against all evil'. More, it was good for gout, something self-evident because fullers who used urine in the preparation and bleaching of cloth, and whose feet were perpetually soaked in it, never suffered from gout. A man's urine was good to drink as an antidote to poison or the bite of an adder and should be used to bathe parts stung by sea adders, scorpions or dragons. A dog bite could be treated successfully with dog's urine, which if mixed with saltpetre could also cleanse scales, scurfy and scabies and cure fretting sores on the private parts. If mixed with the rind of a pomegranate it expelled worms out of the ears. Recorde writes page after page of sage advice such as the foregoing, so it is as well to remember that these are not the writings of a deranged mind, as we might suspect of similar writings today, but the considered advice of one of the foremost scholars and highly educated doctors of his day. In an age before antibiotics, before there was any sort of pain relief or knowledge of the true cause of any affliction, when people often bore open sores on their bodies for weeks and months on end, we should not wonder that the disquisitions of Recorde were prized and valued by everyone who laid hands on his book. Nor is it surprising that such a long medical training was required to absorb so much knowledge, erroneous though we now know it to be. We should be thankful that we live in a more enlightened age.

The years following Wolfe's publication of his medical text in 1547 should have been good ones for Recorde, but indirect evidence suggests they were not. His medical practice seemed not to prosper, and we need to look at circumstances which might support this assertion. A salient point is that he never became a member of the Royal College of Physicians. The inception of the college in 1518 was due to Cardinal Wolsey, who, during an epidemic of plague, granted six physicians a charter to establish a corporation for the promotion of the 'learned art

OF VRINE 63.

But as I haue alwayes sayd, you shall vse them to the counsell of sum learned Physicion: for there is great difference both of the greefe, & of the medicyns.

Medicyns for the stone, both in the raynes and bladder,

Astra Bacca.
Ameos.
Sower Almondes.
Angle toches food
Betony.
Bryony roote.
Bylgrumme.
Chamamell.
Capers barke, namely of the roote.
Carret sede.
Clotte sede.
Docke roote.
Fenel: sede, & roote.
Gotys blood.
Gladyan.
Gromell.
Gumme of Plum tree, & Chery tree.
A hedge Sparow.
Harebell.
Kneholme roote and beryes.
Madder roote.
Hygh Malowes sede & roote.
Mogworte.
Perseley.
Pellyter of spayn
Pyony beryes, which ar black
Radyche.
Sampere.
S. Johns worte
Sperage.
Scholme.
Swynes Fenell.
Sothernewood sede.
Tente worte.
Tutsan

FIGURE 9 Recorde's list of medicines for kidney stones

Medicines for the stone, both in the raynes (kidneys) and the bladder. Whilst some, such as Betony, Dock Root and Radish might be found in a modern-day herbal, others such as Gotys (Goat's) Blood and a Hedge Sparrow certainly would not. Recorde would have expected readers to present the list to an apothecary skilled in transforming these ingredients into medicines. (*The Urinal of Physick* (1547), p. 63.)

of physic'. The charter was confirmed by statute in 1523, which declared that since physicians ought to be 'profound, sad and discreet, groundly learned and deeply studied in physic', the incorporation of the college would ensure compliance with those requirements. Henceforward it had the right to restrict medical practice in London to members of the college and by 1541, shortly before Recorde arrived in London, it began harassing and suing anyone who dared to practice medicine without its licence. Recorde, with his seemingly impeccable academic qualifications and long years of medical study, should have been welcomed with open arms by the august members of the college, but it seems it was not so. Perhaps the small irregularity of his doctorate, previously mentioned, first aroused their enmity, and if so Recorde's medical treatise, written in English, would have done nothing to lessen their animosity. The medical establishment of the time was firmly of the opinion that Latin was the proper language for learned discussion and the very idea that the mysteries and secrets of medicine should be disclosed to all in the vernacular was anathema to their introspective perceptions. Recorde, of course, would have vehemently disagreed, since his whole ethos as a teacher was to inform and enlighten as much as was possible.

The Physicians' Act of 1540 defined medicine as including surgery, and it gave physicians the right to practice surgery when and where they liked, although surgeons were not granted the right to practice medicine. It is quite possible that Recorde took advantage of this provision, since he had a declared interest in anatomy and Bullein later included him in a list of eminent surgeons. In *The Urinal of Physick* he laments the lack of 'an exact book drawn of Anatomy', saying that 'which thing I have long minded, so I intend shortly to accomplish with goodly pictures aptly framed'. It was a book he was never to write, but associating himself with the surgeons, and perhaps even practising their skills, seen by most physicians as a mere craft suitable for barbers and butchers but not learned and educated men, could have done nothing to endear him to their society. It is an anomalous fact that Recorde dedicated *The Urinal of Physick*, a medical work, to the Wardens and Company of the Surgeons in London. Perhaps he intended it as a deliberate snub to the College of Physicians who would not admit him to their ranks.

If Recorde's medical practice was not thriving because he was being harassed by the physicians, it goes some way to explain why he so readily threw it all up and directed his energies and great learning to new fields of endeavour, as will shortly be related. In retrospect, however, we can clearly see that this Tudor MD rose far above the absurd physician-surgeon schism that bedevilled his times.

7
ANTIQUARIAN AND MATHEMATICIAN

Recorde's home in St Katherine Coleman was only a brisk walk along Cannon Street from St Paul's Churchyard and we may suppose that he was a frequent visitor to the Wolfe printing house at the Brazen Serpent. Regarding himself, as he had written in the *Grounde*, as one of the learned who had taken pains to do things for the aid of the unlearned, he must have been pleased to hear from Wolfe that his medical textbook, *The Urinal of Physic*, had been passed for printing by the censors of the Privy Council. Wolfe would no doubt have pointed out to him the words *Cum privilegio ad imprimendum solum* printed on the title page, words which might have puzzled Recorde as much as they do modern scholars. The exact meaning of these words, in Tudor Latin, is disputed, but they seem to mean that by royal privilege or patent the printer has been granted sole, or exclusive, printing rights. Additionally, following a proclamation by Henry VIII in 1538, they also appear to imply rights reserved by the Crown to recall the work after publication, should it be discovered that any treasonable, seditious or heretical matter had been overlooked when the work was first approved.[1] Wolfe, to avoid trouble and make sure that he was well within the law, repeated the words in the colophon at the end of the book.

It is not unlikely that Recorde, a confirmed bachelor far from his own kindred in Wales, came to look upon Wolfe's household as a surrogate family. Wolfe had remarried and, as a close relationship developed between the two men, Recorde might have sat occasionally at Wolfe's

FIGURE 10 A typical doctor of the Tudor period

Wearing a cap and gown, a sign of his academic status, he scrutinises the contents of a urine flask and holds in his right hand an astrolabe. This is an astronomical instrument for determining the conjunctions of planets, a necessary accoutrement for a doctor forecasting the critical days of an illness, the climactic period during which the patient will either recover or die. (*The Urinal of Physick* (1547), title page.)

table alongside his second wife Joan, the rest of the family, other visitors and literary cognoscenti and Wolfe's apprentices and journeymen. The Wolfe family was a large one, and it is tempting to speculate that the eldest son, Robert, was named after his father's distinguished friend.[2] A younger son, Henry, certainly followed Recorde's example and was sent off to university at a tender age. Two daughters, Mary and Sara, were married to the brothers John and Luke Harrison respectively, both printers with their own printing houses nearby. Another daughter, Susan, was married to John Hun, for whom Wolfe had managed to find some space on the premises where he could carry on his haberdashery business. Yet another daughter, Elizabeth, married to Steven Nevenson, a lawyer, occupied a rent-free tenement, also within the environs of the Brazen Serpent. Another family member was Joan's niece, Magdalene Rigthorn, who was 'diseased in her eyes'. This may have been a case of conjunctivitis, the so-called 'pink eye' that is easily treatable today. One wonders if Recorde recommended to her the advice from his medical text, namely that the urine of a child under fourteen years of age, if mixed with honey in a brass vessel, would cure 'the web and the tey in the eye'.

It would not be surprising if Recorde found such a busy and lively environment stimulating, especially when Wolfe's brother Garret paid a visit and technical discussions about the art and craft of printing flowed freely among the masters, workmen and visitors. Wolfe himself may not have put in many appearances during one particular summer, when he was busily engaged in superintending the removal of more than a thousand cartloads of the bones of the dead from the charnel house of St Paul's, conveying them for reburial to Finsbury Fields.[3] He had purchased the chapel from Henry VIII and would in time use it to enlarge his premises, which eventually comprised an extended frontage of houses, tenements, shops, cellars and attics.

A frequent visitor to the Brazen Serpent was the antiquarian John Leland, whose sad fate it was to lose his reason and be certified insane. He died, still mentally deranged, in Wolfe's house in April 1552. Recorde may have met him previously at Oxford, where he also was an elected fellow of All Souls, but it was in all likelihood at the Brazen

Serpent that Leland revealed to Recorde a truly remarkable story. In 1531 he had been appointed chaplain and librarian to Henry VIII, being preferred by the king because of his acknowledged expertise in antiquities and manuscripts. Two years later Henry entrusted Leland with a commission which authorised him to examine and use the libraries of all religious houses in England. Leland spent the following years travelling from house to house, for the most part shortly before their dissolution, compiling numerous lists of significant or unusual books in their libraries. Leland was later to describe how the royal libraries in Henry's palaces at Greenwich, Hampton Court and Westminster were adapted to accommodate the hundreds of books previously kept in monastic collections. What must have astonished Recorde, however, was Leland's revelation that he had discovered many beautiful illuminated manuscripts, together with more mundane collections of writings, apparently archived and forgotten for centuries by the monks who had them in their care, and which no one could read. In fact, these documents were written in Old English, the language of the Anglo-Saxons who had ruled England from about the fifth century until the coming of the Normans in the eleventh.

By the beginning of the sixteenth century, Old English was a long dead and forgotten language and yet today Anglo-Saxon studies are part of the university curriculum. Many people read, write and speak Old English, not only within academia but also among the lay population, who regard learning it as a 'foreign' language a stimulating pastime. Since Recorde's day generations of scholars have painstakingly reconstructed its grammar, vocabulary and spoken sound by analogy with Old German and the modern languages of the Scandinavian countries, although it is a moot point whether Old English as spoken now would have been understood on the streets of King Alfred's Winchester. Nevertheless, the fact that the language is recovered and thrives is entirely due to three men; John Leland, Robert Talbot and Robert Recorde. It is thanks to their pioneering efforts, later emulated by many others, that the Anglo-Saxon manuscripts, upon which all subsequent study has ultimately depended, were rescued from loss or destruction.

Leland was an accomplished linguist, not only skilled in Greek and Latin, but also in 'British', or as we would say today, Celtic or Gaelic, and he was a proficient Welsh speaker. It is tempting to speculate that he and Recorde might have practised their Welsh upon each other. Leland also professed knowledge of 'Saxonish', and it was this claim that excited Recorde's curiosity. It is possible that it was Leland who introduced Recorde to Talbot, a churchman who would later become a prebendary of Norwich Cathedral, and who also had keen antiquarian interests. Once alerted by Leland, Talbot made every effort to track down Old English materials, and many manuscripts passed through his hands containing the gospels, homilies, ecclesiastical and legal texts, grammars and glossaries, as well as Anglo-Saxon copies of such authors as Ælfric, Æthelwold, Orosius and the Venerable Bede.

Recorde too began to collect manuscripts, exchanging them from time to time with Talbot. Both men freely added their own notes and comments on documents that were over five hundred years old when they handled them, and are now five hundred years older, treasured beyond price by academics and many of them highly valued as works of art by an enlightened public. Most are nowadays in the archives of the Parker Library at Corpus Christi College, Cambridge, among them one in particular that provides undisputed evidence of Recorde's previous ownership. Catalogued as Manuscript 138, it contains an abridged version of the *Chronica majora* by Matthew Paris, translated from Latin into Old English, to which Recorde has supplemented the account for the years 449–871 by adding in the margins twenty-five extracts from two other Old English texts. The first of these was the *Anglo-Saxon Chronicle*, the second a West Saxon genealogy that prefaced an Old English translation of Bede's *Historia ecclsiastica*. Both these latter manuscripts are known to have passed through Talbot's hands, and it seems that Recorde was able to borrow from him whenever he wished. Recorde left to posterity over 1300 printed pages in his books but these marginal notes, written in Old English, are the only specimens we have of his handwriting, other than his signature.

The question naturally arises, how did Talbot and Recorde set about recovering and teaching themselves this long-forgotten language?[4]

It seems that Leland's interest waned, perhaps because of the onset of his illness, but Talbot's enthusiasm was undiminished and Recorde emulated his methods of scholarly research. Their ability to make any progress at all ultimately depended on the fortuitous circumstances engendered by no less a personage than King Alfred himself. Alfred, far back in the ninth century, lamented what he saw as a decline in learning, due in no small part to devastating Viking incursions. He began a programme of reconstruction, gathering together in England as many copies of the great works of antiquity and contemporary scholarship as he could obtain from all quarters of Europe. These, of course, were all written in Latin, and Alfred undertook an ambitious programme of translation into Anglo-Saxon, the vernacular language of his people. Alfred himself is supposed to have personally translated some texts, but the work was left mainly in the hands of monastic scribes. With Alfred's encouragement, to help vernacular readers understand Latin, the monks often glossed their texts, that is, they wrote the original Latin between the lines that had been translated into Old English.

These were the clues that Talbot and Recorde needed to unlock the secrets of the forgotten Anglo-Saxon language. They began by studying the vernacular versions side by side with the glosses in Latin, a language they well understood, and then compiling lists of Old English words and defining their meanings by reference to their Latin equivalents. The function of the Old English translations, originally made on Alfred's orders to assist Anglo-Saxons who had difficulty with Latin, was thus completely reversed, with the Latin serving as the key to the meaning of the Old English. Generations of scholars would follow the pioneering efforts of Talbot and Recorde, their joint erudition eventually resulting in the triumphant recovery of Old English that we enjoy today.

Recorde's antiquarian studies were not by any means confined to the Anglo-Saxon period. Proficient in reading and writing Latin, he was equally adept in the study of ancient Greek texts. Perhaps the best indication of the astonishing variety and richness of his interests up to this point in his life is provided by the contents of his library. A contemporary, John Bale, a churchman whose unhappy disposition and habit of quarrelling earned him the unflattering nickname of 'Bilious Bale', developed and published a very extensive list of the works of British authors down to his own times. Despite his fractious and prickly nature, Bale must have been a tolerated visitor to Recorde's house in St Katherine Coleman. He drew extensively on Recorde's collection of manuscripts and books to compile his bibliographies, and in the process left evidence of what a zealous antiquary Recorde was, as well as a valuable listing of the works he found in his library.[5] From this listing we can glimpse the truly amazing range of Recorde's intellectual pursuits, and we can only marvel at the time he must have spent in reading and reflective study in order to absorb so much knowledge in so many different disciplines and then disseminate it in his own books and in his teaching. The subjects are many and diverse, including mathematics and medicine as might be expected, but also surgery, religion and theology, law, astronomy and cosmography, geography, histories of Britain and Ireland and chronicles of the Roman emperors, as well as more esoteric material on alchemy, prophesy and heraldry.[6]

On top of all this study, and in the time allowed by his medical practice, Recorde continued to write. At the end of the *Urinal of Physick* he stated that he had written another book, with the same title but in Latin, 'as more mete for learned ears, and for them that need more precise judgement'. We have no reason to doubt his word on this, although the book, if ever printed, has not survived. Perhaps it was intended as a sop to the College of Physicians.

Wolfe must have been pleased when Recorde presented him with another book which he could profitably print without controversy and which appeared in 1551 with the *cum privilegio* clause attached to the printer's colophon on the last page, indicating approval by the censors. This was the second of Recorde's mathematical textbooks, this time on geometry. He entitled it *The Pathway to Knowledge* (hereafter referred to as the *Pathway*). Its longer title was once again surely due to Wolfe and his keen sense of salesmanship: *The Pathway to Knowledge, containing the first principles of Geometrie, as they may moste aptly be applied unto practise, bothe for use of instruments geometricall, and astronomicall and also for projection of platts in everye kinde, and therfore much necessary for all sorts of men*. This book has subsequently been much criticised by mathematicians as being a less rigorous and poor copy of Euclid's *Elements*. Euclid, of course, was the classical Greek writer whose book on geometry is one of the most influential works in the history of mathematics, serving without interruption as the main textbook for teaching geometry from the time of its appearance around 300 BCE until the early twentieth century, a huge span of time. This, however, is to misunderstand Recorde's purpose. He was writing for practical men who could not read Latin, and who in any case would find great difficulty in following Euclid's complex geometrical premises. Recorde was perfectly able to translate Euclid quite precisely had he wanted to, but his book, written in English, was altogether simpler to understand. The most obvious difference is that Euclid proved his results but Recorde does not. An instance is when he explains a simple method for dividing the arch (an arc) of a circle into two equal parts, by using a square to draw a straight line up from the midway point of a line (the chord) drawn between the ends of the arch:

> Divide the chord of that line into two equal portions, and then from the middle point erect a plumb line, and it shall part the arch in the middle.

Recorde then proceeds to show 'how to do the same thing another way yet', thereby confirming that he intended his treatise for the instruction of practical men as well as those with a more scholarly interest in mathematics.

> If so be it you have an arch of such greatness, that your square will not suffice thereto, as the arch of a bridge or of a house or window, then may you do this. Measure underneath the arch where the middle of his chord will be, and there set a mark. Then take a long line with a plummet, and hold the line in such a place of the arch, that the plummet do hang justly over the middle of the chord, that you did divide before, and then the line doth show you the middle of the arch ... And to the intent that my plummet shall

GEOMETRICALL.

The xxxiii. Theoreme.

In all right anguled triangles, the square of that side whiche lieth againſt the right angle, is equall to the .ij. ſquares of both the other ſides.

Example.

A.B.C. *is a triangle, hauing a ryght angle in* B. *Wherfore it foloweth, that the ſquare of* A.C, (*whiche is the ſide that lyeth agaynſt the right angle*) *ſhall be as muche as the two ſquares of* A.B. *and* B.C. *which are the other .ij. ſides. By the ſquare of any lyne, you muſte vnderſtande a fi=gure made iuſte ſquare, ha=uyng all his iiij. ſydes equall* to *that line, whereof it is the ſquare, ſo is* A.C.F, *the ſquare of* A.C. *Lykewaiſ* A.B.D. *is the ſquare of* A.B. *And* B.C.E. *is the ſquare of* B.C. *Now by the numbre of the diuiſions in eche of theſe ſquares, may you perceaue not onely what the ſquare of any line is called, but alſo that the theoreme is true, and expreſſed playnly bothe by lines and numbre. For as you ſee, the greatter ſquare (that is* A.C.F.) *hath fiue diuiſions on eche ſyde, all equall togyther, and thoſe in the whole ſquare are twenty and fiue. Nowe in the leſt ſquare, whiche is* A.B.D. *there are but .iij. of thoſe diuiſions in one ſyde, and that yeldeth nyne in the whole. So lykeways you ſee in the meane ſquare* A.C.E. *in every ſyde .iiij. partes, whiche in the whole amount vnto ſixtene. Nowe adde togyther all the partes of the two leſſer ſquares, that is to ſaye, ſixtene and nyne, and you perceyue that they make twnety and fiue, why=che is an equall numbre to the ſumme of the greatter ſquare.*

By

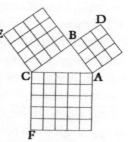

FIGURE 11 Recorde's exposition of Pythagoras's Theorem

He does not prove the theorem, but merely gives a demonstration of its truth, which is sufficient for his intended readership. This may be contrasted with the mathematical proof found in Book I, Proposition 47, of Euclid's *Elements*, which is rigorous but much more difficult for non-mathematicians to understand. (*The Pathway to Knowledge* (1551), 33rd theorem.)

point more justly, I do make it sharp at the nether end, and so may I trust this work for certain.[7]

As he set out to introduce Euclidean geometry to English readers, Recorde found that the English language at that time did not contain many technical terms. Therefore he had two choices; he could either use longstanding Latin or Greek words and hope that they would become familiar, or he could invent new English words. He chose the latter course. So, for example, an equilateral triangle he called a threelike, and a parallelogram a likejamme, unless it had all four sides equal, in which case it was a likeside. Similarly, acute, right and obtuse angles become in Recorde's terminology, sharp, square and blunt angles respectively. Altogether he coined over fifty new words, all of which were the result of a thoughtful endeavour to anglicise mathematical language, with the laudable aim of minimising difficulties for those learning the subject for the first time through the medium of English rather than Latin. Unfortunately, Recorde's terms did not survive the passage of time and consequently, to this day, schoolchildren in geometry lessons have to wrestle with difficult Latin words such as tangent, instead of his much more homely and easily understood 'touch line'.

Recorde himself, of course, was thoroughly familiar with both Latin and Greek editions of Euclid's *Elements*, so much so that his mathematical contemporary John Dee records him as having translated the text into English. If he did, he would have preceded Henry Billingsley, usually acknowledged as the first translator of Euclid, by many years. Unfortunately, no copy of Recorde's alleged translation has survived, depriving him of an honour that might, in some small way, have helped to preserve his memory in later times.

Recorde dedicated the *Pathway* to Edward VI, the young and devoutly Protestant son of Henry VIII, who had succeeded his father in 1547. It must have seemed an astute move by Recorde and Wolfe to align themselves with the new king and the Protestant faction surrounding the throne, most notably the king's uncle and guardian Edward Seymour, duke of Somerset. Soon to be known as the lord protector, Somerset was effectively the ruler of England during the king's minority.

Edward was a precocious scholar who found Recorde's book greatly to his liking, possessing a much-used copy and an astronomical brass quadrant designed by his tutor John Cheke, to be used with the *Pathway* in his mathematical studies.[8]

Recorde had intended that the *Pathway* should consist of four parts and Wolfe printed and included a descriptive contents page headed 'the arguments of the four books'. However, when the book appeared it contained only two of the four and, in his preface dedicated to Edward, Recorde promised not only to:

> put forth the other two books which should have been set forth with these two, if misfortune had not hindered it, but also I will set forth other books of more exacter art, both in the Latin tongue and also in English, whereof part be already written ... and the residue shall be ended with all possible speed.

It is likely that Recorde was referring here to a book he intended to call *The Gate of Knowledge*, but no copy of this work is known to exist. In the preface to the second part of the *Pathway* Recorde described an ambitious and truly astonishing list of works he intended to produce in the near future.

First was what he called *The Second Part of Arithmetic*, a continuation of the *Grounde* but taking arithmetic beyond an elementary level and introducing algebra. For its time, this was an esoteric subject, probably considered by non-mathematicians unfamiliar with its arcane symbolism as akin to cabalistic sorcery. He would in time finish this book but it would be left to someone other than Wolfe to print it. He promised a book dealing with the celestial sphere, the terrestrial globe and the art of navigation, which he also finished although not exactly as first intended, and which Wolfe printed, as will be discussed shortly. He mentioned a work on the art of measuring, using an astronomer's staff and a new type of quadrant that he had invented but about which, sadly, no information has survived. He promised a work on dialling, that is making sundials for use during the day and dials for use at night by reference to the moon and the stars. It is probable that these matters were included in a work he intended to call *The Treasure of Knowledge*, which possibly only ever existed in manuscript; certainly no printed copy is known. He pledged to produce Euclid in four parts, which bears out Dee's assertion that he was indeed the first to translate that seminal work into English. Most astounding of all, considering the tremendous breadth of knowledge required and the time needed to put it into effect, he said he had books 'partly ended' covering such subjects as the 'Peregrination of man', the 'Origination of all nations', the 'State of times and mutations of realms', the 'Image of a perfect commonwealth', the 'Wonderful works and effects in beasts, plants and minerals', and diverse other works in the natural sciences.[9] However, before he could accomplish these wonderful plans, the course of his life changed dramatically. We might wonder if the misfortune which hindered him from completing the last two parts of the *Pathway* was in fact the first intimation that he was about to be drawn, slowly but inexorably, into the dangerous and murky world of Tudor religion and politics.

8

NO MEAN DIVINE

About the month of July 1548 a sudden rumour that King Edward was dead quickly spread through the streets and taverns of London, although Edward was in fact alive and well at Hampton Court. This was a serious matter, since news of the death of a reigning monarch invariably caused unrest among troublesome factions, if not outright rebellion by any number of usurpers and spurious claimants to the throne. Lord Protector Somerset acted swiftly to quash the rumour, instructing the Mayor of London, Sir John Gresham, to apprehend forthwith those responsible for raising the false report. Somerset's messenger to Gresham was Edward Underhill, a trusted member of the band of gentlemen-at-arms originally formed as the personal bodyguard of Henry VIII. Underhill already had a notion about the likely culprits, suspecting the ringleader to be a Norfolk man named Richard Allen. He urged Gresham to provide some officers to accompany him, thinking it likely that he would find Allen in Paul's Walk, the notorious haunt of rumour-mongers and troublemakers.

This proved to be the case, and Allen was arrested and escorted to his house in search of evidence. Here Underhill discovered papers with the king's birth date scrawled on them, followed by calculations forecasting the date of his death. Underhill reported that the 'foolish wretch thought himself so sure about his predictions that he and his papist followers had bruited the king's death all over London'. Allen and his books of conjurations, circles and many other things belonging to his 'devilish art' were unceremoniously hauled before Protector Somerset. Allen protested that the casting of figures to tell things yet to come and to prognosticate the length of men's lives was a lawful science. Somerset would have none of it. 'Thou foolish knave,' he is reported to have said, 'if thou, and all that be of thy science, can tell me what I shall do tomorrow, I will give thee all I have.'

Allen was committed as a prisoner to the Tower, and Underhill rounded up his associates, among them Thomas Robins, otherwise known as Little Morgan, and his brother Great Morgan, an infamous 'dicer' (gambler), together with a shady lawyer called Gascon. This gang of rogues had chambers in different parts of the city where lost or stolen articles were traded, gamblers made their matches, and the gullible went to have their fortunes told. It was also alleged that Gascon and his friends among the young lawyers of the Temple lured the unsuspecting wives of citizens to these chambers in order to debauch them. Somerset wrote a letter to Sir John Markham, the lieutenant of the Tower, instructing him to have these men imprisoned and Allen examined by such as were learned.

Markham, described as being both wise and zealous in the Lord, interrogated Allen and was surprised by the assured manner in which he affirmed that his knowledge came from consultation with the stars. He claimed to know more of the science of astronomy than all the learned men within the universities of Oxford and Cambridge put together, although he admitted he understood no part of the Latin tongue. It was obvious to Markham that someone with great learning and an impeccable reputation among the populace was needed to undermine Allen's certitude and destroy his credibility. It was part of Markham's job to keep his ear close to the ground and he knew that just such a person of note was residing only a few streets away from the Tower. Underhill was despatched forthwith to fetch Dr Recorde.

It did not take Recorde long to expose Allen for the charlatan that he was.[1] He reported to Markham that he 'knew not the rules of astronomy, and was a very unlearned ass'. Furthermore, Allen had imparted to Recorde some esoteric fantasies supposedly resulting from his confabulations with the stars which made Recorde suspect him of being a sorcerer, 'for the which', he said, 'he was worthy of hanging'. This seems an uncharacteristic remark from the mild scholar, but perhaps he was genuinely affronted that Allen dared to claim the scientific knowledge that he himself had laboured so long and so hard to acquire. Recorde very effectively discredited Allen, who became a figure of ridicule rather than a threat to civil order, although he survived his imprisonment to live on into the reign of Queen Mary. The significance of this episode for Recorde was that it brought him to the notice of Lord Protector Somerset and the members of the Privy Council. Hereafter he was drawn increasingly into the duke's administration and his future involvement with the government was to rise and fall with Somerset's fortunes. Recorde's questioning of Allen also occasioned his first meeting with Underhill, after which the two men became good friends. He could not have foreseen that one day this friendship would put his personal safety at great risk.

Underhill was a zealous Protestant whose evangelical fervency had earned him a reputation as a 'hot gospeller'. Only the year before meeting Recorde he had narrowly escaped death at the hands of outraged Catholic women who conspired to have him murdered after he entered the parish church of Stratford-le-Bow and removed from the altar the forbidden pyx, a receptacle in which wafers for the Mass were kept. Long before that, during the reign of Henry VIII, he had single-handedly apprehended the vicar of Stepney and carried him off to appear before the archbishop of Canterbury on a charge that he had hindered the preaching of the gospel in his own church. Recorde was clearly influenced by the example of this young man, who was not afraid to speak out or act against the Catholics in the cause of his Protestant tenets. Gradually relaxing the great caution he had exercised in religious matters since his first bruising experience at Oxford, Recorde too began to speak out on matters of doctrine and faith.

There is some slight evidence that Recorde obtained a doctorate in theology whilst at Cambridge. The Grace Book containing the records of the university for the years 1342–1589 mentions him in connection with the granting of his MD under the heading 'In regard to Doctors in Theology'. It is possible that he became known in the pulpit, speaking perhaps as a lay preacher in his parish church of St Katherine Coleman, which fitted well not only with his religious convictions, but also with his fervent desire to teach and educate his fellow men. Underhill, who knew an effective preacher when he heard one, said he was 'no mean divine', presumably meaning that his Protestant credentials were soundly orthodox and his sermons models of their kind.[2] It is interesting to note that when writing the preface to *The Urinal of Physick*, Recorde was already sniping at the Catholics. Seeking an analogy to show the logical absurdity of those who attacked writing in the vernacular rather than scholarly Latin on the grounds that learning would thereby be brought into contempt, he wrote:

> Many a woman hath been defiled under cloak of confession, and should we therefore refuse the good use that is in Christian confession? Many evil men and heretics have misinterpreted God's word, yet ought God's word nevertheless to be taught vulgarly to all men? Though cardinals and monks have practised to poison men even with the very sacrament of the altar, yet no man will be so mad therefore to eschew the use of that blessed sacrament. Better means it were to set forth publicly all that might do good to the public wealth and straightly to punish the abusers of them, rather than to punish good men and good things because evil men offend.

This was strong stuff, but imprudent, as it was open to any and all interpretations that possible enemies might care to place upon it.

Recorde would have been only too aware of the dangers of publicly committing to one side or the other of the religious divide. On 30 July 1540, shortly before he took up residence in London, the clergyman Thomas Garrard, whom Recorde had known at Oxford and from whom he may have obtained forbidden religious tracts, and who had fled summary arrest by the Oxford authorities, finally met his dreadful end. He, together with Robert Barnes and William Jerome, two other committed Protestant divines who had also offended the Henrician Church, were all drawn on a hurdle from the Tower of London through the streets of the city and across London Bridge to Smithfield. There the three were chained to a single stake and burnt alive, reportedly remaining in the fire without crying out, as quiet and patient as though they felt no pain. At the same time, in that fearful place of execution, Thomas Abell, Richard Featherstone and Edward Powell, all Catholic priests accused of heresy, were hanged, drawn and quartered as traitors. All these men perished in a ruthless demonstration of King Henry's religious even-handedness. Now Henry was dead and his Protestant son Edward was on the throne. This should have spelt safety for outspoken defenders of the Protestant faith such as Underhill and Recorde, but everyone knew that Edward's health was precarious and that his half-sister Mary, a staunch and unswerving Catholic, was next in line to the throne. It was not a good time for Recorde to abandon his customary caution in religious matters and he began to backtrack a little.

Writing the epistle to Edward VI in the introductory part of the *Pathway*, Recorde said that 'as the study of religious matters is most principal, so I leave it for this time to them that can better write of it than I can'. In a later text, writing about astronomy and remarking that the sun, the moon and the stars were ordained by God to serve all nations that are under the heavens, he said:

> Fear not the signs of heaven . . . for oftentimes in the scriptures fear of God is taken for honour of God, and so it is here else otherwise might I answer that the true servants of God which have reposed the love and fear of God in their hearts, are never afraid of any tokens that God sends, but rejoice to see them and glorify God for them. But because in this case there be many divines that can better declare those things than I, which am a man of another profession, I will remit the matter to them.[3]

Caution indeed, and not everyone was enamoured of Recorde's preaching. John Pitts, the Roman Catholic scholar previously quoted, who was not born until two years after Recorde's death, but who presumably read some of his religious writings and sermons that are now lost to us, stated categorically that Recorde overestimated his own competence when he wrote concerning theology. In writing about a subject of which he knew little, Pitts said, Recorde's attempts *merito a Catholocis legi prohibentur*. By this he meant that Recorde wrote about points of doctrine that meritorious Catholics were prohibited by law from engaging in, and his comments probably arise from bitterness against the existing state of affairs. We should not, of course, expect a Catholic to be sympathetic to Recorde's Protestant views, so it was probably generous of him when he went on to say that Recorde's faulty theology should be attributed more to his own limitations and the king's policy than to maliciousness.

Although Recorde had assiduously sought to avoid religious controversy and any suspicion of heresy, he did write on some difficult religious subjects. Among his works listed by Bale are *De Auricula Confessione*, presumably a discourse on the effectiveness or otherwise of confessing sins into the ear of an intermediary priest instead of directly to God by means of private prayer, and *De negotio Eucharistie*, which would seem to have been a discussion about the difficulty of interpreting the Eucharist. Was God's presence during the rite a spiritual reality or was the sacrament merely a symbolic re-enactment of the Last Supper? If the former, exactly how, where and when was Christ present? Did the wafer and wine blessed during the ceremony actually become the body and blood of Christ, or was this also just symbolic? Recorde wrote about these matters in Latin, as befitted such portentous subjects, but it is not known if they were ever printed or simply remained in manuscript. Certainly no printed copies have survived, but in writing them Recorde showed that he was willing to engage in potentially dangerous subjects, albeit with the greatest circumspection. It is likely that, while composing these texts and picking his words and phrases with extreme care, with the flames of the Smithfield fires across the Thames almost literally reflected in his window panes, his thoughts harked back to the iconography that Reyner Wolfe had foolishly allowed to be printed on the title page of the *Grounde*.

The woodcut depiction there has been called by modern scholars 'the quarrel of the abacists and the algorists'. On the left of the picture are two abacists, those who defended the traditional and age-old use of Roman numerals and calculations done with counters on ruled boards. On the right are two algorists, who supported the written methods of calculation done with pen and paper, originally invented in Asia and newly introduced into the west. Since the book teaches calculations with both the pen and with the counters, the illustration was an apt choice, and it is possible that Wolfe obtained it on one of his frequent trips to Frankfurt. If so, he was either remarkably lax in scrutinising it before printing, or he intended to show bravado in allowing it to go forward. Perhaps neither of these alternatives was the case and he may just not have realised the potential for disaster, for himself and for his author, had the censors of the Privy Council picked up on its hidden message. In the picture, on the back wall of the room in which the abacists and the algorists are arguing over a ruled counter and written numerals, are five letters and a date, the date being the year of publication (1543).

The letters VDMIE are more mysterious. They are in fact the initial letters of the Latin phrase *Verbum Domini Manet In Eternum*, which may be loosely translated as 'The word of the Lord endures forever'. This phrase was adopted as a slogan by the Schmalkaldic League, an alliance concluded on 27 February 1531 between several German Protestant princes and cities in opposition to the Catholic Holy Roman Emperor Charles V, who at this time was attempting to suppress Protestantism and the teachings of Martin Luther. In England, Thomas Cromwell, chief adviser of Henry VIII in ecclesiastical matters, was unsuccessful in persuading Henry to join the league. Subsequently, he endeavoured to bring about an alliance between England and the German Protestants by arranging Henry's disastrous marriage to Anne of Cleves. VDMIE became the watchword of the Schmalkaldic League and the rallying cry of Protestantism throughout Germany. Philip the Magnanimous, landgrave of Hesse and leader of the league, had the letters carved on his furniture and emblazoned on the sleeves of his retainers. Many supporters of the Reformation had the letters engraved over their doorways in defiance of the emperor and the pope.

Wolfe, and therefore of course Recorde, were extremely lucky that the censors failed to spot the significance of the letters in the picture. They would hardly have expected such overt religious symbolism in a mathematical textbook and so probably 'rubber-stamped' the book without examining it closely. Had it been brought to the attention of the king he would surely have seen it as seditious, if not heretical. Having deposed the Pope as head of the English church, Henry would have no

FIGURE 12 VDMIE, the cryptic slogan of the Protestant Schmalkaldic League

The letters can be seen above the date on the back wall of the room in which abacists and algorists are arguing over the best method of performing calculations. By printing this iconography, Reyner Wolfe imperilled himself and his author Robert Recorde, putting them in danger of accusations of sedition and heresy. (*The Grounde of Artes* (1543), title page.)

truck with Lutheranism and was not about to have the German monk fill the pope's place in English ecclesiastical affairs. When things went wrong between the king and the Protestant states of Germany, Henry had no hesitation in cutting off Cromwell's head and anyone under the slightest suspicion of questioning Henry's judgements could have expected no mercy. Interestingly, the woodcut was used again on the title page of the 1558 edition of the *Grounde*. By this time both Henry and his successor Edward were dead. Henry's daughter, the staunch Catholic Mary, was the reigning monarch and during her reign it is said that more than three hundred Protestant heretics were burnt at the stake. Perhaps it is no surprise then that VDMIE had disappeared from the woodcut in this later edition.

None of this is to deny that Recorde's devotion to mathematics was evidenced through his identification of number with the Deity. For Recorde, as he explained at the end of the preface in the *Grounde*, God was the:

> true fountain of perfect number, which wrought the whole world by number and measure; he is trinity in unity and unity in trinity, to whom be all praise, honour and glory. Amen.

At the time when Recorde was struggling unsuccessfully to complete the missing two parts of the *Pathway* and was also deeply absorbed in his ambitious programme of scientific study and religious writing, he did finally become irrevocably involved in the deadly religious controversies of the Reformation. His involvement was brought about indirectly by Stephen Gardiner, Bishop of Winchester. Gardiner had remained a moderate churchman while Henry VIII was still alive, renouncing the jurisdiction of the Roman church and favouring the king's divorce from Catherine of Aragon, but had become an outspoken critic of religious policy under Edward VI. He was violently opposed to married clergy, stoutly maintained the real presence in the Mass and argued strongly that there should be no more changes in the established religion during the minority of Edward. This did not at all suit the Protestant coterie headed by the lord protector, nor indeed the devout Edward himself, and Gardiner was swiftly brought to heel. The Privy Council ordered him to preach a sermon on St Peter's day, 29 June 1548, on nineteen articles of faith, thereby causing him either to show his obedience in matters of doctrine, or by his statements to give cause for him to be deprived of his see as a preliminary to further punishment.

Recorde, along with a number of other notable personages and 'such an audience as the like whereof hath not lightly been seen', was present at the sermon, but whether of his own volition or by order of the Privy Council is unclear. It was obviously necessary to have witnesses to Gardiner's utterances, and duly noted was the attendance of:

> Master Robert Recorde, doctor of physic, who was present at the sermon from the beginning to the ending, in such place as he might well hear and understand the said bishop, and give attentive ear unto his preaching.[4]

This seems to imply that Recorde was 'planted' in a favourable spot amongst the hushed audience as Gardiner sailed close to the wind and, contrary to his instructions to show a conformity with the king's proceedings in religious matters, spoke of the sacrifice of the Mass and censured other preachers who spoke against it. In the view of the Privy Council, many of whom were present, he showed himself 'an open great offender and a very seditious man', and the next day he was imprisoned in the Tower of London. Recorde's involvement in the affair was not over, however, although it would be several years before he was called upon as an impartial witness at Gardiner's eventual trial. Meanwhile, the course of his life was about to undergo a drastic change, in a way that he could have little suspected only a short while earlier.

9

COMPTROLLER OF THE KING'S MINTS

Early in 1549 London and the south were rocked by a great scandal. Sir William Sharington, under treasurer of the king's mint at Bristol, was sent to the Tower on charges of fraud and embezzlement. According to an act of attainder, a legal device by which he was condemned without trial, Sharington had illegally coined over £12,000 in testoons, an early form of the more familiar shilling, all minted without the king's warrant. Furthermore, he had made over £4,000 profit for himself by making his coins too light, that is without sufficient content of precious metals, and had then covered his tracks by compiling false records and burning the originals. Matters of coinage were of great concern to the whole populace at this time, but nowhere more so than in the mercantile metropolis of London. At the time of Edward's accession to the throne, his father's military adventures and profligate extravagances had left the Crown almost bankrupt. To remedy this situation Henry had begun to debase the coinage and Protector Somerset had seen fit to continue this dubious practice after his death.

Debasement was possible because a coin actually had two values, its intrinsic worth, which reflected the amount of gold or silver it contained, and its face value, which is what the government said it was worth. Progressively reducing the amount of precious metals in the coins, while at the same time declaring that their spending value was unchanged, provided the Crown with so much additional revenue that, at the time of Edward's succession, income from debasement had become greater than all taxation and the sale of Crown lands put together. Merchants, however, who were shrewd at weighing coins, were not so easily fooled, and to obtain the true weight in gold and silver for their merchandise they began to demand more coins, that is, they increased prices. A practical result was that the cost of basic commodities like bread trebled over a very short time, and year by year everything was becoming more and more expensive. No wonder then that people were baying for Sharington's blood when his misdeeds were exposed.

The charges against Sharington were serious and after his imprisonment he broke down and made a free confession of all his peculations. He acknowledged himself worthy of death or other grievous punishment and on his knees, it is said with a most woeful heart, begged for mercy. While all this was happening Recorde was far away from London, back in his native Wales, and as yet unaware of how the breaking Sharington scandal would affect his future. He had enough troubles of his own to contend with, having just had his first brush with Sir William Herbert, an influential courtier and a notorious soldier of fortune with a dubious reputation. It would not be the last confrontation Recorde would have with this ruthless man, who with hindsight seems such an unlikely and unevenly matched antagonist for the university intellectual.

Recorde was in Wales at the behest of Protector Somerset, who had by now come to appreciate his many talents and great learning. Seeking to ameliorate the impoverishment of the Crown, deposits of iron, lead and silver were actively being sought at this time, and there was a known source of good quality haematite at Pentyrch, near Cardiff. The Pentyrch mine was Crown property, descending by right of inheritance to Henry VIII from Jasper Tudor, but it had recently been granted to Herbert, making him the landlord of the lessees actively carrying on the mining operations there. The area boasted plentiful woodland for fuel and limestone for flux, and waterpower was available to assist iron manufacturing. Altogether the site was ripe for development, either as a bloomery producing wrought iron, or as a blast furnace turning out cast iron.[1]

With his extensive – albeit theoretical rather than practical – knowledge of minerals and mining practices, Recorde must have seemed an ideal choice to manage the site. His lack of experience in the superintendence of any sort of workforce was obviously not seen as a hindrance to his appointment. Perhaps it was thought that his Welsh ancestry would compensate by creating an empathy with the miners and surface workers at Pentyrch. Whatever the circumstances, for a short time Recorde became a *de facto* iron master. His readiness to undertake this task adds to the suspicion that his medical practice in London was not prospering.

Not everyone welcomed Recorde's appointment. A certain Garret Harman had been nominated overseer of all the king's mines by Henry in 1543 and he was probably irked by Recorde's sudden appearance on the scene. He was to reveal himself at a future time as an enemy shamefully determined on Recorde's downfall. On 5 February 1549 an unpleasant altercation with Herbert added to Recorde's troubles, and ultimately engendered a bitter antagonism between the two men.[2] Dubbed in his youth with the acid soubriquet 'Black Will', Herbert lived up to his reputation and his unexpected arrival at the Pentyrch iron mine can only be described as a raid. According to Recorde's later account, the workers were driven off by Herbert's retainers, damage was done to mine property at a cost to the king of £2,000 and iron billets each worth £10 that were Crown property were seized and carried off.

Herbert's motives for these actions are not clear, beyond his obvious desire to enrich himself by whatever means possible. It is conceivable that he thought to bully Recorde into handing over monies arising from the operations at Pentyrch which were due to the Crown and then to intimidate him into keeping quiet about the affair. If so, he chose the wrong man. Thoroughly enmeshed himself in the corruption and malevolence that typified Tudor government and politics, he must have been confounded to encounter a guileless man ready to defend what he saw as the king's best interests. Seeds of bitterness and rancour were sown at Pentyrch that developed into a long-running dispute and eventually into a deadly quarrel.

Recorde's tenure as master at Pentyrch was not a long one. Perhaps to his surprise, and before he left Wales to return to London, he learnt that as of 29 January 1549 he was officially in possession of a Crown appointment, that of comptroller of the newly created Durham House mint in the Strand. In this he was a beneficiary of an experiment begun in the latter years of Henry VIII's reign and continued under Edward VI, the setting up new mints in places remote from the Tower of London, the traditional home of the Royal Mint. Already there were branch mints in Southwark, Canterbury and York, as well as the Bristol mint that William Sharington had embroiled in corruption and public scandal.

Recorde was more than qualified for the job; his scholarly endeavours had included a meticulous study of coinage, and in the *Grounde* he had written a masterly treatise on adding and subtracting the values of gold and silver coins of different denominations.[3] He had shown a great understanding of foreign currency and of the rates of exchange, for example between English money and that of France and Flanders. Furthermore, he was recognised by his contemporaries for his specialist knowledge in the science of metallurgy, evidenced by his assignment to the ironworks at Pentyrch. In a later edition of the *Grounde* not published until 1573, after his death but obviously in Wolfe's hands long before, he replaced the familiar preface to the first and subsequent editions with an address to the young King Edward. In it he wrote of having almost completed a work about the 'rates of alloys... with other mysteries of mint matters'. However, he was well aware of the sensitivity of the subject, touching as it did on the contentious issue of the amount of precious metal in the coinage. Cautiously he declared that he had:

> omitted for just considerations till I may offer them first unto your Majesty, to weigh them, as to your Highness shall seem good; for many things in them are not to be published without the knowledge and approbation of your Highness.

Recorde hardly had time to draw breath upon learning of his appointment at Durham House before the Sharington scandal directly impacted upon his fortunes. As the newly installed comptroller of that mint, it was perhaps not surprising that he was included as one of the four commissioners appointed by Seymour and the Privy Council

FIGURE 13 Sir William Sharington, Undertreasurer of the Bristol Mint
An embezzler and a corrupt administrator, Sharington's downfall led to Robert Recorde's appointment first as Comptroller, and later as Undertreasurer, of the Bristol Mint. Lacock Abbey, Sharington's home in Wiltshire, was confiscated, as were 1,937 ounces of plate, gold and silver estimated to have made up to £14,000 in coin. This treasure may have passed through Recorde's hands while he was in charge at the mint.

and headed by Sir Thomas Chamberlain to investigate the running of the Bristol mint. As a result of these investigations, Recorde and the other commissioners were soon in possession of damaging, not to say dangerous, information that reflected very badly on the lord protector himself. What had started out as a simple case of fraud now threatened the stability of the government. It was discovered that Sharington had been asked by Seymour if he could mint £10,000, sufficient to keep an army of 10,000 men in the field for a month. Sharington agreed that he could, so long as sufficient bullion was available. Presumably he meant by this that the extra sum could be obtained by fraudulent minting activities. Furthermore, he assured Seymour, that so long as there was a mint in Bristol, he 'should lack no money'. Taking possession of the mint and sequestrating the property of the corrupt under treasurer had unwittingly mired Recorde in the dirty business of Tudor politics, but there was no going back.

Thomas Chamberlain was appointed under treasurer to replace the disgraced Sharington and Recorde was made comptroller, in addition to holding that office at Durham House. He took over from the previous comptroller, Roger Wygmore, who was suspected of being involved in Sharington's misdemeanours. Other important staff were changed and Recorde assumed a supervisory role over John Walker, teller, John Mune, provost of the moneyers, Stephen Lathebury, surveyor of the melting house, John Smith, receiver of the testoons (which at that time were being recalled for melting down and reissuing as shillings), and Giles Evenet, graver of irons. Evenet was soon busy engraving new dies for coins bearing the Bristol mintmark TC; (for Thomas Chamberlain; previously the mintmark had been WS for William Sharington). In June 1549, when it was too late for the dies to be changed, Chamberlain was sent off to Denmark as an ambassador and Recorde became under treasurer in his stead. By this accident of timing no coins of the realm ever bore the mintmark RR, to which Recorde's new status probably entitled him, so denying him of another small token of recognition that might have kept his memory alive, at least in the minds of later generations of numismatists.

The comptrollership was an office of profit, giving Recorde an important role in managing the finances of mint operations and requiring him to scrutinise all newly minted coins before issue.[4] His position demanded a great deal of metallurgical knowledge and, although not lacking himself in this science, it must have occurred to him that his brother Richard could be of great help to him at Bristol. At some unknown time, but probably whilst Recorde was in London, their father Thomas died and their mother Ros, now the widow of a wealthy merchant, soon remarried. This may have been the start of a decline in the family fortunes in Tenby and it seems that Richard Recorde began to spend much of his time in an intensive study of alchemy rather than in mercantile activities. The defining objectives of alchemy included the creation of the fabled philosopher's stone, finding the elixir of life that would confer youth and longevity and the ability to transmute base metals into the noble metals, gold and silver. Recorde is unlikely to have had faith in any of these objectives, although he did pursue cursory studies in alchemy, but what must have interested him most was his brother's experience in melting and alloying metals in his search for the secrets of transmutation. Such experience would obviously be of great use to anyone charged with the manufacture of coins. Accordingly, in an age of patronage when it was considered acceptable, if not actually a duty, for persons with power or authority to show preferment to relatives, friends or suitors over all others, Recorde brought his brother to the Bristol mint and installed him in the post of surveyor. The fact that Richard accepted the position and stayed by his brother's side during the next few years is a fair indication that the family business in Tenby was not prospering.

The state of the nation's coinage at this time was dire. The coins most often exchanged in everyday transactions were groats and half groats, and those in circulation were for the most part so bent and battered that they were hardly passable and traders were reluctant to accept them in payment for goods and services. The situation was so bad that the king issued a proclamation decreeing that henceforth:

> all manner of groats, half groats, pence and half pence, not being counterfeits, not being clipped nor fully broken, albeit they might be much crooked, were to be accepted throughout the realm without any manner of refusal or denial.

All mayors, justices of the peace, sheriffs, bailiffs, constables and other officers of the law were instructed to arrest any persons refusing to accept the coins for merchandise, food or exchange and imprison them, to be further punished at the king's pleasure. Clearly, there was need for urgent action to produce new coins, but the problem was a lack of bullion for the manufacturing process. The solution hit upon was to appoint commissioners in every county to survey and make inventories for the king's use of all plate and other valuable objects found in any cathedral or parish church.

It was not long before gold plate from the cathedrals of Wells and Salisbury was passing through Recorde's hands on the way to the Bristol smelting furnaces. Local churches were not spared. On 2 August 1549 Recorde signed a document acknowledging receipt of silver and gold plate surrendered by Roger Walker and Thomas Dole, proctors of the church of St Peter in Bristol.[5] On 12 August William Appowell and Richard Bonde, proctors of St Ewen's, brought in more sacred items to be melted down.[6] Still the treasure poured in; on the very next day Recorde acknowledged receipt of precious plate from William Young and John Pykes, proctors of All Hallows Church.[7] All these receipts, neatly endorsed 'Received by me Robert Recorde, Comptroller of the King his Majesty's mint at Bristol' are still extant, fortuitously affording us today a number of specimens of Recorde's signature.

The articles relinquished by the churches were quite staggering, consisting of chalices, patens, pyxes, crosses both gold and silver, censers, boats for carrying frankincense, tabernacles, spoons of gold, ampullae, candlesticks, cruets, sepulchres, bells, crowns and many other like things. It is difficult to gauge Recorde's feelings as he watched these beautiful, precious and once sacred objects slowly subside into the molten obscurity of the melting cupolas. Perhaps he cared little, viewing the hoard as nothing more than idolatrous remnants of the Catholic church that ought to be put to a better use. Many though, in Bristol and elsewhere, regarded the royal edict as little better than plunder, and it was rumoured that among the plunderers there was no one more corrupt and guilty than the lord protector himself.

In fact, despite the popular suspicions, there is evidence that the churches were paid for the valuables they surrendered. For example, an account for St Peter's records that the proctors, Peter Cowper and

Roger Walker, received £24 3s. 1d for the sale of silver plate to the king's mint at Bristol. Presumably Recorde, as comptroller, was responsible for the assaying of the precious metals and fixing the prices at which they were bought, a particular skill which he was to exercise again in times to come. Under his leadership the quality of the coinage manufactured was very high and the overall performance of the mint could only be considered outstanding. When the financial accounts for the period of Recorde's tenure were examined in 1551 there was found to be a surplus of £218 13s. 2d, a rare and unusual occurrence for any mint at this time when corruption and financial peculation were rife. However, if Recorde felt jubilant at the way things were progressing, his elation was not to last. As had so often happened before in his life, things turned sour at the point of triumph, and he was once more touched by trouble. Again Black Will Herbert came to hound him, this time enmeshing him in a tangled skein of intrigue, cunning and vindictiveness from which he would never escape.

In the summer of 1549, while Recorde and his brother were happily immersed in the pleasant technicalities of minting, there were widespread disturbances in Wiltshire. Protector Somerset, acting without the sanction of the Privy Council, ordered the commander of the king's army, Lord John Russell, to take vigorous action against the rebels. This task Russell placed in the hands of Herbert, who was a skilful and energetic soldier. He was ruthless in putting down the insurrection, so much so that Edward VI wrote in his journal that 'Sir William Herbert did put them down, overrun, and slay them'. However, the rebellion was not over, for by this time the men of Devon and Cornwall were also up in arms. Herbert was instructed to march west with full speed, taking with him 3,000 men from Wales and 2,000 from Gloucestershire and Wiltshire. Such a large army had to be paid and supplies obtained to keep them in the field, and Herbert knew, or thought he knew, where to get the money. He was probably mindful of Sharington's boastful promise to Seymour that he should not want for money so long as there was a mint in Bristol. He probably thought that the new under treasurer would be as pliable and corrupt as the previous office holder, but in this expectation he was mistaken. Recorde refused point blank to divert any funds to Herbert's coffers without the king's sanction, and that authorisation was lacking. Herbert was furious, perhaps understandably, as he had an armed host at his back which, if not paid and fed, was likely to desert and rampage through the west country before skulking home. No doubt he found it hard to believe that a 'nobody' like Recorde would dare, for the second time, to thwart him, one of the most powerful men in the land engaged on the king's business. The repercussions were swift in coming, and Recorde, honest but perhaps unusually lacking in wisdom on this occasion, was ordered back to London. By October 1549 all work had ceased at the mint, Herbert playing a key role in closing it down. Recorde's employment at Durham House was also terminated and that mint too was shut. He was probably stunned when Herbert accused him of treason, the penalty for which, if brought to trial and found guilty, was to be hanged, drawn and quartered.

10

THE MUSCOVY COMPANY

For many years it has been claimed as a matter of fact that Recorde was punished for his lack of cooperation with Herbert and the army at Bristol by being confined to court for sixty days. This is a misunderstanding of the event and its consequences; firstly, in defending the king's interests as he saw them, Recorde had done nothing that merited punishment, although he had undoubtedly upset a few very powerful people and secondly, there was nothing in the statutes of England that prescribed 'confinement' at court, in the sense of fitting and deserved imprisonment, as a punishment for anything. It is more plausible to suppose that Lord Protector Somerset, fearing that Recorde was in possession of facts extremely damaging to his cause, summoned him to court in order to keep him close by his side and prevent him from speaking out of turn. However, events now took a dramatic turn when Somerset was alerted that his rule faced a serious threat.

The rebellions in the west and elsewhere involved a peasant population which had suffered severely through rising inflation and debased currency and was determined to end these grossly unfair practices. When Somerset sent Russell and Herbert to quell the rioters, however, it was without the sanction of the Privy Council, which saw his actions as a step too far in his increasingly authoritarian rule. Sensing a plot intended to overthrow him, Somerset issued a proclamation calling for assistance, took possession of the king's person, and withdrew for safety inside the fortifications of Windsor Castle. Edward, not liking this enforced constraint, wrote in his diary 'methinks I am in prison'. Recorde could well have echoed those sentiments, since he was probably hustled to Windsor too, along with the huge number of retainers, courtiers and hangers-on who invariably followed the court wherever it went.

After publishing details of Somerset's mismanagement of the government, summarised by Edward himself as 'ambition, vainglory, entering into rash wars in mine youth, enriching himself of my treasure, following his own opinion and doing all by his own authority', the Privy Council had him arrested. They brought the king to Richmond and Recorde would have been obliged to follow, since court etiquette absolutely forbade anyone at court to leave without the king's express permission. Although unconfined by prison bars, or even the crossed halberds of sentries, Recorde was probably unable to make a surreptitious exit before the elapse of the reputed 'sixty days'. It may be that he took the opportunity to leave the court during the turmoil following the emergence of John Dudley, duke of Northumberland, as leader of the Privy Council and, in effect, as Somerset's successor. Although Somerset was released from the Tower and restored to the council early in 1550, he was executed for felony in January 1552 after scheming to overthrow Dudley's regime. Edward noted his uncle's death in his *Chronicle*: 'the duke of Somerset had his head cut off upon Tower Hill between eight and nine o'clock this morning'.

One man who would surely have noticed Recorde slipping away from court, if that was in fact what he did, was William Herbert, himself a courtier and therefore often in attendance on the king. No doubt keeping a close eye on his adversary, Herbert was determined that Recorde would not thwart him a third time. He followed his quarry to London and in January 1550 tried once more to browbeat him into handing over gold and silver from the Bristol mint, in addition to the proceeds from the Pentyrch iron mine.[1] Again he was unsuccessful, with Recorde stubbornly refusing to be bullied into handing over what he saw as rightfully belonging to the king and to no one else. It is, of course, extremely unlikely that Recorde would have had any coinage or bullion actually in his possession at this time, and what Herbert probably wanted was some sort of warrant, signed by Recorde in his official capacity as under treasurer and comptroller, giving him a quasi-legal right to the monies.

Armed with such a document, Herbert would have been well able to fend off any challenges to his right to the proceeds.

No doubt wearied by constant and acrimonious argument with Herbert and dismissed, through no fault of his own, from his positions at the mints, Recorde now turned his attention to a preoccupation very much more to his liking. He became involved with the Muscovy Company as a technical adviser and was already writing a book for use by the company's navigators, which he titled *The Castle of Knowledge* (hereafter referred to as the *Castle*). The Muscovy Company was founded in 1551 by Sebastian Cabot, Richard Chancellor and Sir Hugh Willoughby, originally under the name of the Company of Merchant Adventurers to New Lands. Cabot, a much older man than his co-founders, was a famed explorer and cartographer, having already made landfall in Labrador and Nova Scotia. These men persuaded a large group of London merchants to finance a voyage of exploration to find a northerly sea route to China and the Far East – the legendary and fabulous North-east Passage.

Chancellor had recently returned from serving as an apprentice pilot on a voyage to the Levant in the barque *Aucher*. This expedition had been organised by Cabot and was intended to provide much needed experience of long-distance voyaging for English mariners, who were lagging far behind the Spanish and French at this time in the new science of celestial navigation. Willoughby was a renowned soldier as well as a seaman, serving as a captain in the Scottish campaigns and commended by the Privy Council for his adroit negotiations with the Scots. He had just returned to London after campaigning in the border country and eastern marches. Recorde's association with these three men and others he was to meet through their mutual friendship would launch a completely new chapter in his life. However, before he could devote his time to this new interest, another long-standing affair claimed his attention. In December 1550 Stephen Gardiner, Bishop of Winchester, was brought to trial at Lambeth. It proved to be one of the defining show trials of the reign of Edward VI, and Recorde was called to give evidence before the royal commissioners headed by Archbishop Cranmer.

Others who, like Recorde, had been present at Gardiner's allegedly seditious sermon in 1548, were likewise summoned, and accordingly he found himself in very august company indeed. The Privy Council produced Sir Anthony Wingfield, comptroller of the king's majesty's honourable household, Sir William Cecil, the great statesman later to become secretary of state and lord high treasurer, Sir Ralph Sadler, Sir Edward North, the almoner Dr Coxe, Sir Thomas North, Sir George Blage, Sir Thomas Smith, Sir Thomas Challoner, Sir John Cheke, the great educator and tutor to Edward VI, Master Dr Ayre, Master Nicholas Udall and Master Thomas Watson. Recorde, the son of a humble merchant, had come a long way from his Tenby origins, and now the integrity of his testimony before the judicial assembly would have parity with the depositions of some of the greatest men in the land. He was called twelfth in the list of witnesses during the first session of the trial, and his evidence was written down, in Latin, by Sir William Petre, a principal secretary of the council. His deposition begins '*Mr. Robertus Recorde in medicinis doctor etatis xxxviii annorum vel circiter*' (Master Robert Recorde, doctor of physic, of the age of thirty-eight years or thereabouts). Here is a second confirmation of Recorde's birth year being 1512, although the 'thereabouts' shows he was not entirely certain of the year himself. Therefore, the best that can be done is to add *circa* (about) to his birth year (*c.*1512) and to note that the often-quoted *c.*1510 is too early, and arises from confusion about when Recorde actually stated his age. To suppose that he did so in the year 1548, during Gardiner's sermon, and not in 1550, at his trial, is obviously not credible, and yet this misunderstanding is the basis for the constant repetition of the incorrect date.

Recorde was carefully examined about the twelve articles of faith on which Gardiner had been instructed to preach.[2] He testified that the bishop did not admit the authority of the king in his 'tender age' in spiritual matters, saying that he 'was present at the said sermon from the beginning to the ending, in such place as he might well hear and understand the said bishop; and give attentive ear unto his preaching'. If he had admitted the king's supremacy, Recorde continued, he would 'have heard it because he was desirous to hear it said from the bishop's mouth'. Furthermore, in denying the pope's authority, as he was required to do, Gardiner spoke in such ambiguous terms that Recorde and others around him were much offended. Asked if Gardiner had preached that the king 'did godly' in the suppression of monasteries and religious houses, Recorde answered that he did not, 'for if he had, surely he should have heard him and marked it, because he gave himself very studiously to listen to the sermon'.

On the subjects of pilgrimages, relics, shrines, holy bread, holy water, ashes, palms, beads, creeping to the cross, auricular confession, processions and common prayer in English, Recorde testified that Gardiner did not so much as mention them, 'for he surely would have heard them, and noted them, for he purposely went to hear and mark what the bishop would say'. He recalled that the bishop had spoken at length defending the mass, which preaching, Recorde said, 'touched the sacrament of the altar and was an offence to himself and many others'. Following his testimony, Recorde was subjected to a lengthy interrogation by Gardiner's defenders, after which he stood down and his part in the affair was over. It was perhaps inevitable that Gardiner would be found guilty. On 14 February 1551 he was deprived of the see of Winchester and imprisoned in the tower 'without benefit of pen, ink and paper'.

As a contributor to Gardiner's downfall, and in the troublesome religious and political climate of the times, Recorde might well have thought it wise to lie low yet again. He was probably glad to confine himself to his house and return to the pleasant task of completing his book for the Muscovy Company. The *Castle* was to be a treatise on cosmography and the celestial sphere, at a time when there were scarcely a handful of people in England who had any enthusiasm for mathematics or the new methods of navigation. Most mariners knew nothing at all about the subject; twenty years later the great seaman Martin Frobisher was still able to claim that he lacked the wit to understand mathematics or celestial navigation. Many sailors were simply not interested; far into the reign of Elizabeth I the pilot of her majesty's galleon *Leicester* professed 'to give not a fart for cosmography'. It was against this background that Recorde laboured to enlighten seafaring men, and he was at pains to show that he knew his subject. In the *Castle* he wrote:

imagine a ship swift of sail to be at the Cape of Cornwall ready to make sail towards the west directly, and to have a great gale of wind, it is possible that she may run 240 miles in twenty-four hours, for I have been at a trial of a greater course, therefore I speak (as men say) within my bounds.³

In other words, in modern parlance, Recorde was saying, 'I have been there, I have done it and I know what I am talking about'. It would be interesting to know the length of his voyage during the 'trial' and whether he set foot on foreign shores, but alas he fails to tell us. He also clearly understood the use of the stars in navigation, writing that 'The most northerly constellation is the Lesser Bear, called Ursa Minor, and Cynosura, and contains in it seven stars. This is the chief mark whereby mariners govern their course in sailing by night'. Again, in discussing navigational instruments he noted that the great circle representing the horizon was divided into thirty-two parts, 'which do betoken the points of the shipman's compass, or the thirty-two winds notable in sailing', which he then listed, beginning with 'North' and ending with 'West and by the South'. He further demonstrates his maritime knowledge when writing 'by trial of mariners, [the sea] has been found in a few places, a hundred fathoms deep'.

As well as associating with Cabot, Chancellor and Willoughby, Recorde also became acquainted with Anthony Anes Pinteado, a famed Portuguese pilot. Philip Jones, a wealthy merchant who later financed an unsuccessful expedition to find the North-west Passage, recorded some thirty-five years after the event that these men conferred with Recorde at his house to discuss their proposed voyage into the northern seas.⁴ Pinteado, who would later accompany the English seaman Thomas Wyndham on a voyage to Guinea (the name then used for the western part of Africa which now includes Nigeria), had specialist knowledge of Portuguese trading voyages down the coast of Africa to India, and would have been able to impart valuable navigational information to the fledgling Muscovy Company.

It was undoubtedly from Pinteado that Recorde learnt of fabulous places such as Calicut, a port on the Malabar Coast of India which, he said, 'it were as much folly to make a doubt of it, as it were to make a doubt of Babylon or Jerusalem', or the Straits of Magellan, a navigable passage between the Atlantic and Pacific oceans, 'as is well known by the navigations of the Portingales (Portuguese) and Spaniards'.[5] Although Recorde undoubtedly had theoretical knowledge of celestial navigation, it is likely that Chancellor and Willoughby would have tended to listen more attentively to Pinteado, a practical sailor like themselves, expounding on this difficult and esoteric subject. What, then, made these seafaring men listen to what Recorde, an intellectual and at best an amateur seaman, had to say about the elusive North-east Passage? A few lines from another of his books, *The Whetstone of Witte* (hereafter referred to as the *Whetstone*), dedicated to the Muscovy Company, about which more will be said in due course, provides the clue. In the dedication Recorde wrote that:

> no men before you dared attempt [to sail into the uncharted northern seas] since the time of King Alfred his reign. I mean by the space of 700 years. No others before that time had passed that voyage, except only Ohthere, that dwelt in Halgolande, who reported that journey to the noble King Alfred, as it does yet remain in ancient record of the old Saxon tongue.[6]

This intriguing quotation shows that Recorde had read the Old English version of Paulus Orosius's *Historiarum Adversum Paganos Libri VII* (Seven Books of History Against the Pagans). This text was translated into Anglo-Saxon in King Alfred's lifetime and it contained an account of the visit to Alfred's court by the Viking trader Ohthere, who told the king that his home was in Halgolande (today Helgeland in northern Norway). Here he lived 'north-most of all Norwegians, since no one lived north of him'. Ohthere described his travels south to Denmark and north to the White Sea in great detail, and mentioned Sweoland (Sweden), the Sami people (Lapps) of Finland, the Cwenas who lived north of the Swedes and the Beormas whom he found living around the shores of the White Sea. He told Alfred that the waters around his homeland were best for whale-hunting, and presented some walrus tusks to the king, commenting that these animals had 'very noble bones in their teeth'. The seamen of the Muscovy Company must have listened with intense interest and astonishment as Recorde, an academic and the only man in the whole of England able to read the Anglo-Saxon version of Ohthere's travels and realise its value to the planned expedition, supplied the sort of information they so desperately needed.

These rugged weather-beaten sailors, about to risk their lives in search of a shorter sea-route to the orient, venturing where no Europeans, as far as they knew, had ventured before, must have hung on to his every word. One can imagine them pumping Recorde for every scrap of intelligence he was able to impart. Did this Ohthere mention anchorages and safe havens? Were people friendly or likely to be belligerent? Were fogs and storms frequent? Did ice hinder navigation? Where could fresh water be found? Their very survival might depend on the answers to these and a hundred and one other questions. Recorde was certainly in the thick of planning for one of the first of the great Tudor voyages of exploration that would later characterise the reign of Elizabeth I.

On 22 May 1553 three ships sailed from Tilbury in search of a passage to the far east via the northern seas. The largest vessel, the *Edward Bonaventure*, was captained by Chancellor, who was appointed pilot-general of the voyage. Sir Hugh Willoughby was in overall command, sailing in his flagship, the *Bona Esperanza*, and the smallest vessel, the *Bona Confidentia*, brought up the rear. All three ships safely reached the Lofoten Islands, lying within the Arctic Circle, where they were scattered by a storm. The *Bona Esperanza* and the *Bona Confidentia* were able to meet again the following day, but the *Edward Bonaventure* did not rejoin them. The loss of Chancellor, the only expert navigator in the small fleet, was to prove fatal for the crews of the ships separated from him. They sailed on aimlessly, steering by their rudimentary charts, and they became hopelessly lost. Passing the entrance to the White Sea without discerning it and failing to find a safe harbour on the island of Novaya Zemlya, they eventually anchored in the deep estuary of the River Arzina, to the east of present-day Murmansk. Here they were trapped by the winter ice, and Willoughby and all his men perished.

The ships and the frozen bodies of their crews were found the following summer by Russian fishermen.

Meanwhile Chancellor, after waiting a week for the lost ships to rejoin him, pressed on alone. He reached the Russian port of St Nicholas on the Dvina river, where the town's inhabitants, although amazed by the great size of the foreign ship in their harbour, were cautiously friendly. From them Chancellor conceived the idea of travelling overland to the court of the tsar, Ivan the Terrible. He reached Moscow the following winter, having written ahead asking to be met by the tsar's emissaries, and he gained Ivan's agreement to the Muscovy Company's monopoly of the trade between Russia and England. Chancellor arrived back at Harwich in 1554, and finding that Edward VI had died, he delivered a letter of greetings from Ivan into the hands of the new monarch, Queen Mary. Recorde would no doubt have been pleased by Chancellor's safe return, and was probably agog to hear at first hand an account of his travels.

11
THIS TALK DELIGHTS ME MARVELLOUSLY

The Castle of Knowledge comprises four books, or parts, dealing respectively with the material (earthly) and celestial spheres, the making of model spheres, how to use the model spheres for astronomical purposes and lastly an in-depth discussion with added proofs of propositions only briefly mentioned in the previous parts. The last treatise also includes many astronomical tables that, said Recorde, 'are set forth very pleasant and profitable'. It appears that his original intention was to make the book a treatise on navigation, but at some point he changed his mind and shifted the emphasis to astronomy. It is likely that he was influenced in this decision by meeting with practical seamen and pilots, which led him to realise that their rather limited understanding of the heavens precluded any teaching in the arts of celestial navigation until they had acquired more fundamental knowledge. So this great educator once more dipped his quill in his inkpot and set to work to bring enlightenment where previously there had existed only incomprehension.

Recorde returned to the dialogue form of presentation used so successfully in his first book on arithmetic, and in the first treatise we find the master explaining to the scholar the four elements of which all matter is comprised.

MASTER: And these four, that is, earth, water, air and fire, are named the four elements. That is to say the first, simple and original matters, whereof all mixed and compound bodies be made, and into which all shall turn again.

SCHOLAR: Oftentimes have I heard it, that both man and beast are made of earth, and into earth shall return again. But I thought not that they had been made of water, and much less of air or fire.

MASTER: Of earth only, nothing is made but earth, for a herb or tree cannot grow (as all men confess) except it be helped and nourished with air convenient, and due watering, and also have the heat of the sun. And generally, since everything is maintained by his like and is destroyed by his contrary, then if man cannot be maintained without fire, air and water, it must needs appear that he is made of them, as well as of earth, and so likewise all other things that be compound.

The scholar is enthused by the new and strange things he is learning. The master, however, fears that the conversation is straying from the immediate point.

SCHOLAR: This talk delights me marvellously, so that I cannot be weary of it, as long as it shall please you to continue it.

MASTER: This talk is not for this place, partly for that it is more physical then astronomical, and partly because I determined in this first part, to omit the causes and reasons of all things, and briefly to declare the parts of the world, whereof these four elements, being uncompounded of them self, that is simple and unmixed, are accounted as one part of the world, which therefore is called the Elementary part, and because those elements do daily increase and decrease in some parts of them (though not in all parts at once) and are subject to continual corruption, they are distinct from the rest of the world, which hath no such alteration nor corruption, which part is above all the four elements, and compasses them about, and is called the Skye, or Welkin, and also the Heavens. This part hath in it diverse lesser or special parts, named commonly Spheres, as the sphere of the Moon which is lowest, and next unto the elements. Then above it, the sphere of Mercury, and next to it the sphere of Venus, then follows the Sun with his sphere, and then Mars in his order, above him is Jupiter, and above him is Saturn. These seven are named the seven Planets, every one having his sphere by himself severally, and his motion also several, and unlike in time to any other. But above these seven planets, there is another heaven or sky, which commonly is named the Firmament, and has in it an infinite number of stars, whereof it is called the Starry sky. And because it is the eighth in order of the heavens or spheres, it is named also the eighth sphere. This heaven is manifest enough to all men's eyes, so that no man needs to doubt of it, for it is that sky wherein are all those stars that we see, except the five lesser planets, which I did name before, that is Saturn, Jupiter, Mars, Venus and Mercury.

Having expounded at length on the basic tenets of Ptolemaic astronomy, Recorde moves on to the second treatise, systematically instructing the scholar in the making of a solid celestial globe, used to map objects in the sky. He also gives detailed instructions on how to make another 'in ring form with hoops', which a few pages on he correctly names an armillary sphere. It is in the third treatise, however, that Recorde begins to demonstrate not only his great knowledge of geography, but also his understanding of the practical uses of astronomy in the service of maritime navigation. Instructing the scholar in the operation of the globe he has just constructed, he asks him to set it to an arbitrary position and then to read off the degree of latitude indicated.

> MASTER: But now for better practice, set your globe to some other elevation.
>
> SCHOLAR: I trow I have set the pole high enough.
>
> MASTER: Let it stand. What is the number of the elevation?
>
> SCHOLAR: I do see between the pole and the horizon in the meridian diverse numbers, but I take that number only which touches

the horizon, and I take that also of the two orders of numbers which descend from the pole, and that is here now 71.

MASTER: That is the latitude or elevation of the pole at Wardhouse, where our new venturers into Moscovia do touch in their voyage.

It is, of course, a pedagogical pretence on Recorde's part that the scholar's arbitrary reading of the globe should fortuitously coincide with the latitude of Wardhouse. This was the name by which English seamen knew Vardo, a small fishing port in Norway lying on the coast of the Barents Sea. Modern day satellite mapping places the port at a latitude of 70° 22′ 14″ N, so Recorde's figure of 71° was remarkably accurate. He had no doubt given the Willoughby expedition this all-important figure, and was also able to tell the navigators that the port reputedly remained ice-free all year round (due, we now know, to the effects of the warm North Atlantic drift). Because of his extensive reading in English history, Recorde would have been well aware that from the fourteenth century onwards small barques had sailed from Lynn, Yarmouth, Hull, Scarborough and other east coast ports into the northern seas for fishing and buying fish, going as far east as Wardhouse (Vardo) in the time of Henry V. Tragically, despite knowing its latitude, Sir Hugh Willoughby and his companions were unable to find this safe haven after separating from Richard Chancellor.

Recorde amply shows his knowledge of contemporary geography when he teaches the scholar to calculate the circumference of the earth. After performing the necessary arithmetic, the scholar arrives at a figure of 21,600 miles.

SCHOLAR: Whereby I know that 21,600 miles does answer unto 360 degrees in the sky. And so it should seem that those are the just number of miles about the earth.

MASTER: You need to make no doubt thereof, except you doubt whether there be any part of the earth without the circuit of heaven. Or else you doubt whether the earth be in the middle of the world.

> SCHOLAR: The first doubt were too foolish, and for the second (albeit I doubt nothing of it), yet I assure myself by your promise of the full proof thereof in the next treatise.

Unfortunately Recorde was in error, the currently accepted value being about 24,900 miles. However, here we have an example of Recorde practising what he preached, calculating and proving the circumference to his own satisfaction and not merely quoting ancient Greek authorities such as Eratosthenes, who gave an over estimate of about 29,000 miles. He goes on to probe the scholar's understanding of the lessons he is being taught.

> MASTER: You do consider that this conclusion being true, they that dwell 5,400 miles from us, do dwell a quarter of the earth from us?
>
> SCHOLAR: That must needs be so, for four times 5,400 does make the whole circuit of 21,600 miles.
>
> MASTER: And so they that dwell from us any manner of way, 10,800 miles, they dwell half the compass of the whole earth from us?
>
> SCHOLAR: It follows so by the former reason.
>
> MASTER: It is well known by the navigations of the Portuguese and the Spaniards, that there is almost south from us, certain places inhabited about 6,300 miles, as namely at the Strait of Magellan. Also at the great foreland of Africa, commonly called the Cape of Good Hope, are there diverse regions replenished with inhabitants, and they be from us southward above 5,200 miles. Then northward we have good knowledge of diverse countries beyond us above 1,200 miles, which both joined together, do make from the great foreland of Africa aforesaid in the south, unto Wardhouse in the north part of Norway, about 6,400 miles, which is more than a quarter of the compass of the earth.

It seems that at this point Recorde was forced by circumstances which will become apparent in due course to lay the book aside for a time. The third treatise concludes with a rather cryptic exchange between the two interlocutors.

> MASTER: Therefore now shall you depart for a time... and at your return, I will instruct you more exactly in all the premises, and other diverse conclusions, which now I have omitted of purpose.
>
> SCHOLAR: I am most earnestly bound unto you for your great gentleness, which I pray God to requite, since I cannot and who will else I know not.
>
> MASTER: Farewell then, and remember your own profit.
>
> SCHOLAR: The author of all profit, continue and increase your profit, that you may have quiet time to travail for the profit of many.

These lines, for the first time in Recorde's writing, betray the increasing anxiety disturbing his mind. He was probably still being harassed by the Royal College of Physicians, so that his medical practice was not prospering and income from it was, at best, small. His employment by the Crown was now ended, but Sir William Herbert persisted in his animosity and malevolence and continued to harry him. Above all, the lines show that Recorde had little expectation of his books bringing him any substantial financial reward, and pecuniary difficulties may have worried him. Little wonder then that his concerns eventually weighed so heavily on his shoulders that he was unable to keep these personal matters out of his writing.

Recorde would eventually complete the *Castle* whilst under very great strain. Although it is impossible to be certain, he probably had a draft to hand which he polished and finished long after he wrote the first words. When he did take up his pen again to begin the fourth treatise, it is not difficult to imagine him recalling happier days at the house in London of which he was so proud. The scholar returns to his studies, and the opening exchange between him and the master is quite delightful, although it will become apparent in due course that the dialogue is loaded with meaning.[1]

FIGURE 14 The Castle of Knowledge, atop of which sits Ptolemy

It was a common misunderstanding in medieval times to confuse Claudius Ptolemy of Alexandria, the renowned astronomer and geographer, with the one-time ruling Ptolemy dynasty in Egypt, hence the crown on Ptolemy's head. William Moreton had the supporters in this image, Destiny and Fortune, recreated in plaster at Little Moreton Hall, his home in Cheshire. (*The Castle of Knowledge* (1556), title page.)

SCHOLAR: If the inexplicable benefit of knowledge did not enforce me to forget all bashfulness, I might think it too much shame so often to trouble my master from his earnest studies, and to stay him from his profitable travail with my importune craving of knowledge, namely since I cannot recompense any part of his pains. Yet his gentleness is such, that he seeks more the profit of others than his own pleasure or peculiar commodity, and therefore will I boldly enter into his house. Are you at home sir?

MASTER: I am always at home for my friends, if I be not with them from home. Yet sometimes I cannot be at home for myself.

SCHOLAR: The less for me and such as I am, that often trouble you more for our own commodity, than for your gain.

MASTER: I seek to gain no more than competently may serve my necessary uses, with convenient regard to my charges. But if I offend any ways in coveting money, I assure you it is to bear the charges in setting forth such monuments of knowledge as were marvellously profitable for all men, very pleasant to many men, and yet esteemed only of wise men. But since I cannot do the good that I would, and others want will which have goods in excess, I must do as many others do, wish good to all men and help them as I can.

However, Recorde put his troubles aside and settled down to the interesting business of explaining and proving to the scholar the fundamentals of Ptolemaic astronomy; that the earth was in the centre of the universe and remained quite stationary, while the sun and the planets and the moon and the stars all moved in their daily motions around the earth. But even as he composed the *Castle* with its centuries-old explanations of the heavens, he must have received news from the continent that surprised him. Although none of Recorde's papers or letters have survived, it is inconceivable that this great scholar, able to read and write Latin with ease, did not correspond and exchange information with associates and counterparts on the continent. Reyner Wolfe had many acquaintances throughout Europe to whom Recorde could have been introduced. He perhaps utilised the agency of Simon Grynaeus, the German scholar who visited London, to set in motion a growing list of mathematical, scientific and theological respondents who might have wanted, in the spirit of the age, to exchange information with him. Furthermore, as a university man and a senior academic, he would surely have had many opportunities to converse by letter with his counterparts in foreign institutions. Since we do not know this with certainty, we can only hope that someday a letter or two from him will surface somewhere in Europe to confirm this supposition.

The astonishing news that Recorde received from the continent, by one means or another, was that a rather obscure Polish cleric by the name of Nicolaus Copernicus, who like himself had studied medicine, mathematics and astronomy, had formulated a model of the universe that placed the sun rather than the earth at its centre. This contradicted everything Recorde had written in the *Castle* and had been at such pains to teach the scholar. We can imagine his perplexity as he wrestled with the momentous implications of the Copernican theory and struggled to understand it. It could not have been long though before his brilliant mind discerned the truth, and it appears he quickly accepted the theory in its entirety. Now the problem for Recorde was what to do about his Ptolemaic book? It was too late to rewrite it. Wolfe was probably waiting for the manuscript, and in any case it would take many months of study, even years, before he could write a competent book on a sun-centered universe. However, it seems that Recorde's academic pride prevented him from simply ignoring the thesis of Copernicus as so many others did when they first learnt of it. He had at least to mention it, and so, right in the middle of the *Castle*, we find some extraordinary paragraphs. Recorde begins by having the scholar offer a comment concerning the proof given to him, that the earth is in the middle and centre of the world.

> SCHOLAR: I perceive it well, for if the earth were always out of the centre of the world ... absurdities would at all times appear.

Historians of astronomy regard the next exchange between master and scholar as the first time in the English language that the

Copernican thesis was placed before the reading public, albeit only briefly. Interestingly, it also shows that Recorde was well aware that the ancient Greek astronomer Aristarchus had hypothesised that the sun was actually at the centre of the universe, with the earth revolving around it, centuries before Copernicus advanced his theories.

> MASTER: That is truly to be gathered, howbeit Copernicus, a man of great learning, of much experience, and of wonderful diligence in observation, has renewed the opinion of Aristarchus Samius and affirms that the earth not only moves circularly about its own centre, but also may be, yes and is, continually out of the precise centre of the world thirty-eight hundred thousand miles. But because the understanding of that controversy depends on profounder knowledge than in this introduction, I will let it pass until some other time.
>
> SCHOLAR: Nay sir, in good faith, I desire not to hear such vain fantasies so far against common reason, and repugnant to the consent of all the learned multitude of writers, and therefore let it pass for ever and a day longer.
>
> MASTER: You are too young to be a good judge in so great a matter. It passes far your learning, and theirs also that are much better learned than you are, to improve his supposition by good arguments, and therefore you were best to condemn nothing that you do not well understand. But another time, as I said, I will so declare his supposition that you shall not only wonder to hear it, but also peradventure to be as earnest then to credit it, as you are now to condemn it. In the mean season let us proceed forward in our former order.[2]

Satisfied that he had let his readers know the true state of affairs, Recorde must have gone ahead with the *Castle* knowing that it had been completely outdated by the theories of Copernicus. Nevertheless, an understanding of the Ptolemaic system continued to be of great use to navigators, as it was right up until modern times and the advent of satellite navigation systems, so the text was far from redundant. It was a great success when published by Reyner Wolfe in 1556. Trouble still clung to Recorde, however, as we may adduce from the final exchange in the book.

> MASTER: But farewell for a time. I am driven to omit teaching of astronomy and must of force go learn some law.
>
> SCHOLAR: The God that is author of true astronomy, and made all the heavens for men to behold, keep you in health and clear from all trouble, that you may, as you mind, accomplish your works and finish well and speedily the fruits of your study.
>
> MASTER: Amen, and Amen.

12

PEDAGOGUE AND POET

Although overworked and deeply troubled at the start of the 1550s, Recorde was obliged to seek commissions among the London printers in order to supplement his income. According to Anthony à Wood he translated several books from French, but in this Wood was misled by his informant William Cole. It seems that Recorde's name was strangely confused by Cole with the verb 'record', capitalised and spelt with a final 'e', in a poem by Robert Copland celebrating the work of the translator Andrew Chertsey.[1] This led to the erroneous assumption by Cole that Recorde and not Chertsey was the translator in question, so sadly we cannot add mastery of the French language to Recorde's many other achievements. More certain is the fact that he was employed by the printer John Kyngston to collate the first and third editions of Robert Fabyan's *New Chronicles of England and France*, originally published by Richard Pynson in 1516.

Robert Fabyan was a wealthy London merchant with an amateur inclination to study and write about history. Just why Kyngston thought his book worth reprinting is unclear, although it must have had some merit since two more editions had followed the first. He may simply have been mischievous, knowing that Cardinal Wolsey had commanded many copies of the *Chronicles* to be burnt because it made too ample a disclosure of the excessive revenues of the clergy. In the event Kyngston's brief to Recorde was to compare Fabyan's efforts with the chronicle *Historia Regum Britanniae* (History of the Kings of Britain) written by the Welsh cleric Geoffrey of Monmouth in the twelfth century, a work widely popular in its day and regarded as credible well into the sixteenth century. Kyngston's aim seems to have been to produce a much improved fourth edition of Fabyan's *Chronicles*, which eventually appeared in 1559 described as 'newly perused'. The 'peruser' was, of course, Robert Recorde.

Adhering to his strong pedagogical instincts, Recorde was not content merely to edit the book; where he thought it lacking, he included additional information of his own. These additions supply further proof of his great learning and scholarship, and the immense amount of reading and study he must have undertaken. For example, where Fabyan, in discussing the Early British kingdoms, says that 'great discord was in the land which grieved the people sore under five kings' without identifying any of them, Recorde supplies their names in a footnote:

> The five kings that be omitted here are found in certain old pedigrees, and although their names be much corrupted in diverse copies, yet these are the most agreeable: Rudaucus king of Wales, Clotenus king of Cornwall, Pinnor king of Loegria [the Celtic province containing most of southern England except Cornwall], Staterius king of Scotland, Yeuan king of Northumberland.[2]

Even more substantial contributions to the *Chronicles* follow, which clearly signpost Recorde's extensive and esoteric knowledge of British and English history. Fabyan describes how Emerianus, king of the Britons, was deposed because of his cruelty.

> And after him reigned twenty kings successively, the one after the other, of the which ... is no mention made either for their rudeness, or else cruelty, or discordant means, or manners used in the time of their reigns, the which disorder clerks disdained to write or put in memory. After them came Bledgaret, succeeded by another nine kings of whom for the former consideration is neither name nor time of reign put in memory.

Fabyan was mistaken about what was known of these twenty-nine kings, for Recorde proceeded not only to name every one of them, but also to relate their dynastic relationships, as well as giving some account of their characters and notable events of their reigns.[3]

Reyner Wolfe too was intent at this time on printing his own chronicle, derived from jottings left to him by John Leland. They comprised a mass of notes concerning British antiquity, gathered by Leland on his extensive travels around England. It is an intriguing speculation that Recorde might have had a hand in editing some of this material for his friend, a task well suited to his talents. In the end Wolfe employed a young man named Raphael Holinshed to put together the work, which eventually appeared posthumously in 1577. Titled *Holinshed's Chronicle*, the book left unacknowledged the researches of Leland and the many years of collating by Wolfe, leaving Holinshed to garner the fame. The work is still in print.

Recorde's writings give many instances of his knowledge not only of history but also of contemporary matters. For example, in the preface to the *Pathway* he is at pains to assert that science, not magic, is the art by which many things are discovered and invented, promising to write at some future time:

> of such pleasant inventions, declaring what they were, but also will teach how a great number of them were wrought, that they may be practised in this time also. Whereby shall be plainly perceived, that many things seem impossible to be done, which by art may very well be wrought. And when they be wrought, and the reason thereof not understood, than say the vulgar people, that those things are done by necromancy. And hereof came it that Friar Bacon was accounted so great a necromancer, which never used that art (by any conjecture that I can find) but was in geometry and other mathematical sciences so expert, that he could do by them such things as were wonderful in the sight of most people.
>
> Great talk there is of a glass that he made in Oxford, in which men might see things that were done in other places, and that was judged to be done by power of evil spirits. But I know the reason of it to be good and natural, and to be wrought by geometry (since

perspective is a part of it) and to stand as well with reason as to see your face in common glass.

Historians of physics now discount the discovery of the telescope by Galileo Galilei and usually assign the invention to Johann Lippershey, a German-Dutch spectacle-maker, at the beginning of the seventeenth century. However, current scholarship suggests that the proto-scientist Roger Bacon, a member of the order of Friars Minor, may well have possessed a telescope in the thirteenth century, and Recorde, writing in the middle of the sixteenth, obviously did not doubt that he had such an instrument.

Although Recorde was always quick to dismiss superstition and the role of magic in science, he was not loath to make use of astrology and alchemy, both today regarded as pseudo-sciences, when it suited his purpose. For example, consulting the juxtaposition of the stars was necessary in medicine to determine critical days, the point in the progression of an illness when the malady would begin to triumph and the patient would probably die, or alternatively when natural processes would bring about the patient's recovery.

Recorde, of course, was well aware of his brother's interest and experiments in alchemy, and in addition to utilising his skills in smelting metals at the Bristol mint may well have assisted him in his alchemical studies. Another of Recorde's acquaintances was Richard Eden, who was also known as an alchemist. Eden, later to achieve distinction for his translations of Spanish and Portuguese cosmographical and navigational works for the Muscovy Company, was thus known to both the Recorde brothers. At a dinner given by Sir John Markham, the same lieutenant of the Tower who had sent Edward Underhill to fetch Recorde to interview the false prophet Allen, Eden met Richard Whalley, Recorde's old friend from his Cambridge days. Whalley offered Eden a salary of twenty pounds a year plus board for himself, his wife and a manservant, if he would devote himself to the search for the quintessence, the means by which life could be prolonged. Whether Eden seriously believed that he could find the secret in six months, as he promised, or was just stringing his gullible patron along, is difficult to say. Whatever the case,

Eden and Whalley soon fell to violent quarrelling about the details of their contract and eventually parted on ill terms.

The alchemical experiments of the two Richards – Recorde and Eden – are memorialised in an old poem preserved by Elias Ashmole in his *Theatrum Chemicum Britannicum*, published in 1652. Verse sixteen of this poem, entitled *Bloomfield's Blossoms, or the Camp of Philosophy*, by William Bloomfield, attests in rather poor rhyme that:

> Then brought they in the Vicar of Malden
> With his green Lion that most royal secret,
> Richard Recorde, and little Master Eden,
> Their metals by corrosives to Calcine and Fret;
> Hugh Oldcastle and Sir Robert Green with them met.
> Roasting and boiling all things out of kind,
> And like Foolosophers left off with loss in the end.

It is a moot point whether these alchemists, and Richard Recorde in particular, heeded his brother's resolutely practical advice that one might 'search all secret knowledge and hid mysteries by the aid of number', that is, by the application of mathematics.

It is likely that the two brothers were much in each other's company at this time, perhaps in London, but equally possibly back in Tenby, as Recorde owned property in the town. On one occasion he had to defend his landed interests in the law courts when a neighbouring weaver and tanner repeatedly allowed his animals to stray on to his property, causing a great deal of damage.[4] Recorde certainly retained connections with his native Wales, an instance being his long association with Robert Lougher, a clergyman and lawyer. Lougher, like Recorde, was born in Tenby, and he also became a member of All Souls College, Oxford, in 1553, two years after Recorde's election. Lougher married Elizabeth, granddaughter of John Rastall the printer, in Tenby, and after Recorde's death was admitted to Doctors' Commons, later becoming Regius Professor of Civil Law at Oxford. It may well have been to Lougher that Recorde turned when he found himself in sore need of legal advice in the troublesome times ahead.

Recorde's brother Richard also married an Elizabeth, the daughter of William and Joan Baenam of Tenby, by whom he had three sons and five daughters, his eldest son and heir being named Robert, presumably after his renowned uncle.

One of Recorde's attributes which has been almost completely neglected by writers and scholars in the past is that he was a competent poet. W. F. Sedgwick said that he often indulged in very passable poetry in his published books. We may deduce that he read poetry from a comment in the *Castle* where, after giving the Greek and Latin names of the zodiac, he says the names 'may be Englished as I have under written, and are often times mentioned of our English poets'. The first example we have of his rhyme is a few lines of doggerel on the reverse of the title page of the *Grounde* (Recorde uses the archaic word 'froward' to mean a contrary person, one difficult to deal with):

> To please, or displease, sure I am,
> But not of one sort, to every man.
> To please the best sort would I fain,
> The froward to displease I am certain.

In the preface to the *Grounde*, discussing the differences between men and animals, Recorde says that numbering is 'the chief point (in manner) whereby men differ from all brute beasts. For as in all other things (almost) beasts are partakers with us, so in numbering they differ clean from us'. He then shows his droll humour in the following poem:

> The fox in crafty wit excedeth most men.
> A dog in smelling hath no man his peer.
> To foresight of weather if you look then,
> Many beasts excel man, this is clear.
>
> The wittiness of elephants doth letters attain,
> But what cunning doth there in the bee remain?
> The emmet foreseeing the hardness of winter,
> Provideth vitals in time of summer.

> The nightingale, the linnet, the thrush, the lark,
> In musical harmony pass many a clerk.
> The hedgehog of astronomy seemeth to know
> And stoppeth his cave where the wind will blow.
>
> The spider in weaving, such art does show,
> No man can him mend, nor follow, I trow.
> When a house will fall, the mice right quick,
> Flee thence before, can man do the like?

In the second verse, Recorde uses the Old English word emmet (æmete), meaning an ant. The last verse was omitted from the first three editions of the *Grounde*, apparently by oversight of Wolfe's compositors, and not added until the fourth edition in 1552. At the end of the book, above a list of errata, Recorde apologises in rhyme for any mistakes readers might encounter:

> No head so headly can be given,
> But error slippery will creep in.
> For man without error scarcely can be,
> So that error exceedeth all diligency.
>
> Patiently therefore I pray you bear,
> Those few faults committed here.
> More pleasant profit I give by reading,
> Than grievous grief by errors often dying.

The title to another piece of doggerel at the beginning of *The Urinal of Physick*, 'An admonition to the readers', leaves no doubt as to the pedagogical purpose of the rhyme:

> Read all, or leave all,
> So am I perfect and steady.
> To read part and leave part,
> Is to pluck the limbs from the body.

In the dedication of this book Recorde laments 'how quick sighted most folk are in other men's acts, and how prone to control, correct and rebuke all men's doings save their own'. He continues by observing that 'methinks a man cannot be too circumspect in avoiding all just causes of reproach and blame. Yet, since there can be nothing well done, but somebody against it will rail and jest, there is no other way to avoid all such taunts, but to live idle, and meddle with nothing, so that I may, as it seems justly, thus conclude':

> Since it is so
> Procured by kind,
> What one can do
> By wit and mind,
> And no other thereto
> Some fault will find,
> Yea less or more
> Shall not him blind.
>
> Better it is,
> Thy pen to refrain,
> Than often this
> To move disdain,
> Thou shall not miss
> But feel some pain,
> If thou love bliss,
> Therefore abstain.

In the preface to the *Pathway* Recorde extols the usefulness of geometry. In a poem remarkable for its scope and breadth concerning occupations and trades far removed from his own professional and academic milieu, he offers the following verse for the reader's consideration:

> Since merchants by ships great riches do win,
> I may with good right at their feat begin.
> The ships on the sea with sail and with oar,

The pathway to KNOWLEDG, CONTAI-

NING THE FIRST PRIN=

ciples of Geometrie, as they
may moste aptly be applied vn=
to practise, bothe for vse of
instrumentes Geome-
tricall, and astrono=
micall and
also for proiection of plattes in euerye
kinde, and therfore much ne=
cessary for all sortes of
men.

Geometries verdicte

All fresshe fine wittes by me are filed,
All grosse dull wittes wishe me exiled:
Thoughe no mannes witte reiect will I,
Yet as they be, I wyll them trye.

FIGURE 15 Geometry's verdict

Recorde poetically declares that geometry will sharpen the wits, but then drolly suggests it will do little for the dull-witted. 'All fresh fine wits by me are filed – All gross dull wits wish me exiled – Though no man's wit reject will I – Yet as they be, I will them try'. The word 'plattes' in the sub-title means 'plans', as in drawings or diagrams. (*The Pathway to Knowledge* (1551), title page.)

Were first found, and still made, by Geometry's law.
Their compass, their card, their pulleys, their anchors,
Were found by the skill of witty Geometers.

To set forth the capstan, and each other part,
Would make a great show of Geometry's art.
Carpenters, carvers, joiners and masons,
Painters and limners with such occupations,
Broiders, goldsmiths, if they be cunning,
Must yield to Geometry thanks for their learning.

The cart and the plough, who doth them well mark,
Are made by good Geometry. And so in the work
Of tailors and shoemakers, in all shapes and fashion,
The work is not praised, if it want proportion.
So weavers by Geometry had their foundation,
Their loom is a frame of strange imagination.

The wheel that doth spin, the stone that doth grind,
The mill that is driven by water or wind,
Are works of Geometry strange in their trade,
Few could them devise, if they were unmade.
And all that is wrought by weight or by measure,
Without proof of Geometry can never be sure.

Clocks that be made the times to divide,
The wittiest invention that ever was spied,
Now that they are common they are not regarded,
The artsman contemned, the work unrewarded.
But if they were scarce, and one for a show,
Made by Geometry, then should men know.

That never was art so wonderful witty,
So needful to man, as is good Geometry.
The first finding out of every good art,

Seemed then unto men so godly a part,
That no recompense might satisfy the finder,
But to make him a god, and honour him forever.

So Ceres and Pallas, and Mercury also,
Eolus and Neptune, and many other mo,
Were honoured as gods, because they did teach,
First tillage and weaving and eloquent speech,
Or winds to observe, the seas to sail over,
They were called gods for their good endeavour,

Then were men more thankful in that golden age,
This iron world now ungrateful in rage,
Will yield them thy reward for travail and pain,
With scandalous reproach, and spiteful disdain.
Yet though other men unthankful will be,
Surveyors have cause to make much of me.

And so have all lords, that lands do possess,
But tenants I fear will like me the less.
Yet do I not wrong but measure all truly,
And yield the full right to every man justly.
Proportion Geometrical hath no man oppressed,
If any be wronged, I wish it redressed.

J. Payne Collier, commenting on the *Castle* in 1866, observed that 'although the work is merely one of science, the author has interspersed verses, some of them of no ordinary excellence'.[5] He points out that the preface opens with a striking quatrain, that is, a stanza of four lines, typically with alternate rhymes, as Recorde's poem exemplifies:

> If reason's reach transcends the sky,
> Why should it then to earth be bound?
> The wit is wronged and led awry,
> If mind be married to the ground.

Collier also pointed out that no one had ever taken notice (so far as he knew) of an admirable hymn contained in the preface, asking his readers to bear in mind at what an early date it was composed. He comments that the hymn is in the same measure as previously, and is precisely of the character and length that would be wished, full of reverence and poetry. 'Here', he says, 'we have force, brevity, grandeur and simplicity, the essentials of good poetry, united with the truest and most comprehensive piety':

> The world is wrought right wondrously,
> Whose parts exceed men's fantasies.
> His maker yet, most marvellously
> Surmounteth more all men's devises.
>
> No eye has seen, no ear has heard
> The least sparks of his majesty.
> All thoughts of hearts are fully bared
> To comprehend his deity.
>
> Oh Lord, who may thy power know?
> What mind can reach thee to behold?
> In heaven above, in earth below
> His presence is, for so he would.
>
> His goodness great, so is his power,
> His wisdom equal with them both.
> No want of will, since every hour
> His grace to show he is not loath.
>
> Behold his power in the sky,
> His wisdom each where does appear.
> His goodness does grace multiply,
> In heaven, in earth, both far and near.

This striking composition was set to music by Dr John Harrison, and in 2010 sung as the Gradual Hymn in a service celebrating the fifth centenary of Robert Recorde's birth. Held in St Mary's church, Tenby, the same church in which he worshipped with his family as a child, his spirit would surely have rejoiced to hear Welsh and English voices uplifted in song and singing his words.

As a concluding example of Recorde's poetry, that on the title page of the *Whetstone* may be quoted, in which he links his first book, the *Grounde*, with his last, by declaiming the essential ability of mathematics to sharpen the mind:

> Though many stones do bear great price,
> The *Whetstone* is for exercise
> As needful, and in work as strange.
> Dull things and hard it will so change,
> And make them sharp, to right good use.
> All artsmen know, they cannot choose,
> But use his help, yet as men see,
> No sharpness seemeth in it to be.
> The *Grounde of Artes* did breed this stone,
> His use is great, and more than one.
> Here if you like your wits to whet,
> Much sharpness thereby shall you get.
> Dull wits hereby do greatly mend,
> Sharpe wits are fined to their full end.
> Now prove, and praise, as you do find,
> And to yourself be not unkind.

Leaving now Recorde's poetry and returning to the fraught year of 1551, we find him struggling to complete the *Castle*. He was worried by penury and his increasing altercations with William Herbert. Like most Protestants, he was probably fretting and fearful about the likelihood of the Catholic Princess Mary succeeding the ailing Edward VI. Then an event occurred which promised to lift at least some of his burdens from his shoulders. In this he was deluded, although it would be some time before he realised it.

13

SURVEYOR OF THE MINES AND MONIES

In May 1551 Recorde was recalled to government service when he was appointed surveyor of the mines and monies in Ireland. Mining and its products had long been a concern of the Crown and under Henry VIII there were attempts to locate and work metalliferous mines. Of particular interest was the mining of lead, a valuable commodity in its own right, but also because galena, the ore from which lead was smelted, often contained substantial deposits of silver. Between the years 1541–5 the lord deputy of Ireland had written to Henry several times advising him of the great value of mineral wealth in Ireland, specifically mentioning lead. Henry seems to have taken a long time to wake up to the prospect of obtaining an indigenous source of silver, but once alerted he acted decisively. The English at this time were greatly impressed with the superior skill of the Germans in mining and metallurgy. Accordingly, in 1546 Henry commissioned Joachim Gundelfinger, a German mining captain from Augsburg, to recruit skilled miners from among his countrymen and bring them to Ireland.

Gundelfinger was directed to work closely with Garret Harman, the man who had been irked by Recorde's unexpected appearance at Pentyrch. Gundelfinger, however, did not have a very high opinion of Harman and in a letter to Henry expressed doubts about his knowledge of mining and extraction metallurgy. The seeds of discord were thus sown, even before the lord deputy and his council were authorised to form a commission for mines in September 1546, when the king issued a prest of £1,000 to allow mining activities to begin. Henry died the following year and matters went into abeyance while the Privy Council and the country were preoccupied by the coronation of Edward VI and the nomination of Somerset as lord protector.

Only after the four tumultuous years that followed, culminating in the fall of Somerset and his condemnation to death, and with the duke of Northumberland now firmly in control, did attention return to the Irish silver mines. In 1550 Edward sent instructions to Sir James Croft, the newly appointed lord deputy, for the management of the mines at Clonmines on the shores of Bannow Bay, County Wexford. At the same time Gundelfinger was reported to be ready in Antwerp, with a contingent of fifty experienced miners, and a new Irish mint was created in Dublin. Recorde, as yet unaware of these developments, must have been greatly surprised by his appointment in May 1551 as surveyor of the mines and monies in Ireland. By letters patent he was also made inspector-general of the Dublin mint. Both were remunerative positions which, at the very least, promised an alleviation of his pecuniary worries.

Clonmines, originally a port until its shallow harbour silted up, had been mined sporadically for its silver deposits for many years before the Tudor monarchy focused on its potential wealth. The name 'Clonmines' predates any mining activities, and probably originates from the Irish *Cluain-min*, meaning 'smooth meadow'. At the end of June Recorde once more set foot on the deck of a ship, a vessel that carried him and his brother Richard, together with servants and officials, across the Irish Sea. A second ship transported Harman, Gundelfinger and the company of miners. They arrived at different ports but met at Waterford, where the miners tarried while they purchased necessary tools and supplies.[1] Recorde arrived at Clonmines in the company of Sir James Croft, who left him there in July 1551 with instructions to inform him as soon as possible as to 'what profit is like to rise of the mines'. While awaiting the arrival of the miners, Recorde organised a careful scrutiny of the activities of the Dublin mint and its officials.

When the men finally turned up Recorde quickly applied himself to bringing order, discipline and system to the mining operations. He arranged for shifts to be worked so that activities might be continued day and night and within six months a considerable amount of ore had been brought to the surface. Unfortunately, relations between Recorde and the foreign workers deteriorated with alarming rapidity. Gundelfinger, relegated to second in command over his own men, was contemptuous of Recorde's 'book-learning' interference, as he saw it, set against his own years of practical mining experience. A bitter dispute arose between the two men concerning the best place to erect a stamping mill and melting house to process the ore. Unable to reach agreement, Recorde seems to have become exasperated, and in his anger called the Germans knaves and drunkards, threatening to bring in miners from Derbyshire who, under his guidance, would mine more ore than the foreigners ever could or would. Gundelfinger retorted that he had come at the king's calling, not only to get the ore, but also to teach the Englishmen the art of mining; he was willing to do so according to his promise, but had not 'come thither to learn of any man'. The contentious buildings were eventually erected at Ross, about eight miles distant from the mines, where sufficient wood for fuel and waterpower was available, although in the end the site pleased no one.

As if this was not strife enough, Garret Harman was deeply resentful that Recorde and not himself had complete authority over the whole Irish mining operation, although his nomination as overseer of all the king's mines had probably lapsed with Henry's death. He peevishly complained about Recorde to the Privy Council, saying that 'following his [Recorde's] own wilful mind it was September before he would cause the erection of the melting house to begin'. To Harman's uncooperative attitude was added the indignation of Martin Pirry, treasurer of the Irish mint, who was ordered to heed Recorde's counsel, and to whom Recorde had been introduced as a superior 'to whom he must declare and make privy the state of the mint ... and what money he had there ... like a true and faithful servant'. Recorde thus found himself confounded by resentment at every turn, and far from his worries being eased, more troubles began settling inexorably upon his shoulders.

That Recorde possessed the necessary theoretical competence to superintend the mines can hardly be doubted. He showed considerable

FIGURE 16 Ore Stamping Mill and Melting House
A water-powered stamping mill of the type which caused the initial discord between Robert Recorde and Joachim Gundelfinger, over the best place for its construction at Clonmines. The stamps crush lead ore before it is melted and silver extracted. A melting house (furnace) can be seen in the background. (Agricola, Georgius, *De Re Mettallica*, Book VIII, 1556.)

knowledge of metals and mining, holding for example that given the colour, taste and other qualities of (underground) water, it was possible to 'tell what veins of earth or metals is in that place'. His possession of practical experience in mining, however, is more doubtful. It could hardly have been otherwise, given his long years in academia and the practice of medicine, not to mention time spent in reading and writing books. His deficiencies in this respect are revealed by an incident that occurred in December 1551, when he learnt of a French whaling ship lying in port at Waterford with barrels of whale oil for sale.[2] Supposing that the use of oil-burning lamps rather than flickering candles would give better illumination in the mines, thus improving output and ameliorating the conditions for the miners, Recorde and his brother set off to buy the oil.

Together they purchased fifteen tons and carted it away without immediate payment, probably by offering the whalers some sort of promissory note on the strength of Recorde's official position. Later, when the money was paid, the Frenchmen were incensed that the coinage proffered was Irish, debased and of much less value than English, French or Spanish currency. They sailed away in high dudgeon, complaining that Recorde had abused his commission and committed plain robbery. This was not the end of the matter, and time would show that this incident may have been the spark that brought the full wrath of the state and the animosity of the Privy Council down on Recorde's head. Nor were the German miners mollified by his purchase on their behalf. They disdainfully rejected his well-meant oil lamps, preferring the traditional tallow candles stuck with a lump of clay to their leather skull-caps. This placed the light right in front of their eyes and left both hands free to wield hammers, chisels and picks. They would have nothing to do with Recorde's lamps which, although much brighter, had to be carried in the hand or placed on the floor, casting long gloomy shadows.

In January 1552, aware of the discord at the mines and worried by spiralling costs, the Privy Council wrote to Recorde, requesting him to report on his stewardship. At the same time, unknown to him, they asked Gundelfinger to report on the work done so far. Recorde was unequivocal, stating that the results were utterly incommensurate with the expenses incurred. He reported that the king's charges were running at over £260 per month, while the gains were not above £40 per month, with the result that the Crown was losing £220 per month. The blame for this miserable state of affairs he placed squarely on the heads of Gundelfinger and the German miners, complaining of their waste and incompetence, and bluntly stating that Irish and English miners he had engaged locally were better skilled then ever they were. He also said that there was a good deal of jealousy and squabbling between the nationalities employed.

Gundelfinger was equally scathing in his report. The miners of his company were, he said, 'greatly disquieted and discouraged to continue their service unless they had a better surveyor and overseer than Master Recorde, or else they shall neither be able to serve or to live'. He angrily claimed that Recorde had:

> thus used them to weary them and cause them to depart, to the intent that he only might get the use of the ore and mines into his own hands to enrich himself thereby, and cause the King's majesty to lose the great cost he had done, and the great gain and profit that is now coming, which will be inestimable, the chief charges being already past.

It is clear that relations between the two parties had completely broken down and that neither tact nor diplomacy on Recorde's part could restore them. It is highly probable that he was completely out of his depth when dealing with the fractious men in the mines, who did not speak his language and whose hard and dangerous occupation made them a tight-knit group not amenable to outside intervention.

With Gundelfinger as the sole intermediary between them, the captain could easily bias his countrymen against Recorde if it suited his purpose. The quiet scholar, unused to raising his voice and accustomed to respectful attention when addressing his medical patients or people of his own class, was quite at a loss as to how to restore the situation. The accusations against him were very grave, and with losses running at such a high level the Privy Council decided to appoint a commission of enquiry to determine the reasons and to sift the truth of the whole affair. The commissioners arrived in Ireland in June 1552 and began an exhaustive and searching investigation into all the technical and financial aspects of the Clonmines operation.

Before Recorde could begin to defend his reputation and rebut all charges of wrongdoing, Garret Harman revealed himself to be a pernicious enemy and, in a 'brief certificate' to the Privy Council, he piled on the complaints.[3] When reading his accusations it is as well to remember that we only have his side of the story. Recorde's answers and rebuttals to the charges have not survived, always assuming they were ever written down in the first place. Firstly, Harman alleged that when the captain wanted a bridge speedily erected over a body of water near the mine, which 'would have advantaged the King much ore', and to which the local seneschal contributed 100 marks and ample manpower, Recorde would not cause it to be made'. A second pontoon bridge over a tidal inlet, which Harman said he showed Recorde how to construct by floating it on boats, was also never made, even though 'it would have been a good passage at low water to pass at all times'. The notion that a man whose skills even Gundelfinger questioned could demonstrate and instruct Recorde, the greatest savant of his day, in anything at all, much less simple bridge building, is to be doubted. Harman, however, had more to say. He complained that carts and wagons had been made to carry ore between Clonmines and Ross, but that the road was broken and impassable in one place and Recorde, even though he promised to do so, would not mend it. We do not know if the road was subsequently mended and Harman was just being too impatient. He likewise said that a hoy (a small single-masted coasting vessel), which was used for the transport of supplies and materials between Waterford and Clonmines, 'might have been bought cheap enough but Recorde would not buy it'. Perhaps Recorde saw no need to buy a vessel, presumably owned by a third party, which was already in use for the carriage of their materials. Again, Harman alleged that the lack of a lighter (a barge) much hindered the works, so he bought one himself for forty marks, but it was left idle and never brought into service. By what authority he made this purchase is not clear and presumably he wanted restitution, but possibly Recorde chose not to be involved with unauthorised expenditure. All these grumbles amounted to little more than pettifogging, but they would prove extremely damaging to Recorde's case if taken seriously by the Privy Council.

Harman surely went too far though when he accused Recorde of peculation and profiteering. According to his 'brief certificate' the surveyor, by the king's authority, had purchased food and other necessities on behalf of the miners, which he then resold to them at exorbitant prices. Wheat, for example, which he bought for 3s. 4d a bushel he sold to them for 10s. 8d a bushel. They had to pay seven crowns a barrel for herrings, bought for four crowns a barrel, and for seven score hakes bought for four crowns he made them pay for six score the sum of eight crowns. However, it was never part of the mining contract that the miners should be subsidised, and they were expected to provide their own food. Perhaps Recorde, good steward of the king's money that he was, was buying cheap and selling at whatever price his limited market would bear. It is after all human nature to complain that prices are too high whenever they are higher than one wishes to pay. Likewise in the case of the miners' shoes. Harman said that the 'poor men' had attempted to buy shoes in England at 12d or 13d a pair, because they had heard that there was a scarcity of shoes in Ireland, but were dissuaded by Recorde. Then, when they needed new footwear, Recorde made them pay his own shoemaker 3s. 4d a pair for them. If there is any truth in this story, could it be that Recorde prevented the miners from wasting their money on inferior footwear that he knew would quickly deteriorate and fall apart in the harsh conditions of the mines? We simply do not know how much credence to place on Harman's complaints. The possible explanations suggested above serve merely to show that there could be alternative reasons for Recorde's alleged actions. It should be remembered that Harman was a bitter enemy of Recorde and that the charges are completely uncharacteristic of the physician and scholar, who had exhibited total honesty on many previous occasions.

However, worse was to come, and Recorde must have reeled when another letter from 'Joachim Gundelfinger, captain of the mines at Clonmines, to the Privy Council in England', acrimoniously accused him of being the cause of all their difficulties.[4] The letter began:

> Whereby I understand [from Garret Harman] that your Honours are displeased that silver was not made here. I certify, your Honours, that the fault was not in us the Dutchmen [i. e. the Germans] ... but being here, how we lack victuals although we

pay double the price for such as we receive. Our folk of the mines are fallen sick, and three are dead for lack of victuals, and also we are not paid in due times. In which matters I shall desire your Honours to take order. God be praised the melting and fining house and the stamping mills are ready, so that workmen be set forth and we will make 300 oz. of silver every week, by the love of God, so long as God do send ore to the King's majesty. I and the miners are not come hither to learn of Master Recorde, [but that] if they will follow me and the miners we will do our best... to teach the English nation truly, according to my promises made to my Lord High Treasurer.

This smacks of a face-saving exercise, intent on blaming Recorde for the low or non-existent output of silver, which was the consequence of the miners being starved, even to the point of death, by him. It is surely extraordinary to infer that Recorde, a man steeped in the Hippocratic traditions of medicine, who had spent many years of his life studying the best ways to prolong life, callously stood by and let men die for want of sustenance. Nevertheless, the accusations kept coming. Two bands of soldiers had been sent from England under the command of Captains Randell and Devenish, presumably to keep an eye on the silver and provide a guard during its transportation. Harman spitefully grumbled that Recorde 'took up twenty men from Captain Randell and as many from Captain Devenish and set them to work practising the devices of his own brain, whereof came no profit to his majesty'. Was Harman deliberately being obtuse, given that Recorde may in fact have been helping the two captains by finding employment for their otherwise idle and perhaps bored men? Again, we just do not know the true facts.

What could not be denied was that the mines were a failure and the output of silver, set against the costs of winning it, was abysmally small. Recorde may well have known that the shortcomings were caused by over-expectations at court, based on an assay of silver content in a sample of lead taken from the mines before any work began. This was probably not a random sample, as it should have been, but a carefully selected specimen provided by Gundelfinger and Harman. Although not exactly a criminal act it was a highly dubious one and now Recorde, the innocent party to their duplicity would, as the surveyor and inspector-general in charge, be called to account for it. Because of the seriousness of the charges against him Recorde was finally summoned to London early in 1552. He must have defended himself adequately on this occasion because the Privy Council appears to have found the reports that dammed him to be either false, exaggerated, or unproven, and he was ordered to return to his post in Ireland. Unhappily, his troubles were far from over, and the strife, disagreements and unsatisfactory state of affairs continued unabated, with very little silver being made, for the rest of that year.

14

NEMESIS

Worried and fretting about his future, Recorde must have been disconsolate when matters regarding the silver mines came to a head in the new year. In January 1553 the Privy Council sent letters to Sir Edward North, Sir John Mason, Sir James Croft and Sir Martyn Bowes, asking them 'to examine and consider certain books sent to them touching the state of the mines in Ireland'. They were instructed to question Robert Recorde, Joachim Gundelfinger and such others as they thought fit, not only to gain a better understanding of affairs at the mines, but also to investigate the controversies that had arisen between Recorde and Gundelfinger.[1]

The four noblemen replied the following month, and their findings struck right to the heart of the matter. The king's profit, they reported, was only £474, while his charges amounted to £3,478 and that was not counting £2,000 'before they wrought one day', that is, before work even began. Presumably they were alarmed by the serious charges made against Recorde but, given that their enquiries only lasted a short time, they were unable to reach any definite conclusions regarding the truth or otherwise of the allegations made against him. Nevertheless, their report was damning, and the only possible conclusion was that the mines were a complete failure.

The Privy Council immediately instructed Recorde to 'proceed to his account and cease henceforth to use any more the commission granted to him for the mine matters [and presumably the mint matters as well] in Ireland'. Gundelfinger must have assumed himself vindicated, but an order to close the mines and discharge the German, English and

Irish miners, which came without warning in the middle of March, swiftly persuaded him otherwise. The council directed that they should all be given 'a sum of prest money for their conduction hither', after receipt of all the ore, lead, coal, instruments and other things appertaining to the king's majesty that they had in their custody had been delivered into the care of some 'trusty person'.

Recorde, although returning to England under a cloud, could hardly have expected to find himself now embroiled in a squabble with the French. The whalers from whom he and his brother had purchased the barrels of whale oil in Ireland had complained so vociferously about being robbed of their just dues that the incident reached the ears of the French ambassador. He brought the complaint to the English court and sought to make political capital out of it. Northumberland, the *de facto* head of state, was furious. With Edward VI sickly and not expected to live and the threat of trouble with Spain looming over the probable succession, the last thing he wanted was an altercation with the French. The lord mayor of London was ordered 'to call to him Master Recorde ... and certain expert merchants haunting Flanders ... to know what fifteen tons of whale oil, taken by Recorde of a Frenchman in Ireland, be worth'. In February 1553 a warrant was hurriedly issued to Sir Edmund Peckham to pay 'to John de Reparasso, Frenchman, the sum of £34 pounds in full recompense of certain oils by him delivered in Ireland to Robert Recorde'. Presumably this generous payment, on top of what Recorde had already given, was sufficient to convert 'taken' into 'delivered' and the incident was settled, probably much to his relief, before it got too far out of hand.

What is certain is that Recorde's standing at court and with the Privy Council was now at an all-time low. With an enquiry into the proceedings at the silver mines sure to be set in motion eventually, it is not surprising that anxiety weighed heavily on his mind and his health began to suffer. He could hardly have been cheered when events temporarily conspired to push all thoughts of the silver mines debacle to the back of his mind and bring to the fore what he and all followers of the Protestant faith had long feared.

On 6 July 1553, at the age of fifteen, King Edward VI died from a lung infection, possibly tuberculosis. Anxious to prevent the

Crown going to his half-sister Mary, since he feared that she would restore Catholicism and undo his reforms as well as those of his father Henry VIII, he planned to exclude her from the line of succession. Persuaded by Northumberland to favour Lady Jane Grey, the wife of his son Guildford, Edward named Jane as his successor. Shortly before Edward's death, Mary was summoned to London to visit her dying brother. She was warned, however, that the summons was a pretext and that she would be detained, thereby facilitating Jane's accession to the throne. She fled into East Anglia and from there wrote to the Privy Council giving orders for her proclamation as Edward's successor. On 10 July 1553, the same day her letter to the council arrived in London, Lady Jane was proclaimed queen by Northumberland and his supporters. By 12 July Mary and her supporters had assembled an armed force at Framlingham Castle in Suffolk. Jane was deposed on 19 July and she and Northumberland were imprisoned when the Privy Council changed sides and proclaimed Mary as queen. Mary rode triumphantly into London on 3 August 1553, accompanied by a procession of over 800 nobles and gentlemen.

Mary's first act as queen was to release Stephen Gardiner from imprisonment in the Tower. Later that year she sanctioned the execution of Northumberland and the following year Lady Jane Grey and her husband Guildford followed him to the block. Recorde, living once more in the parish of St Katherine Coleman, might have feared Gardiner's revenge, since his testimony had played a part in the prelate's imprisonment, but the churchman proved to be remarkably forgiving towards his former adversaries. However, as a well-known Protestant intellectual whose problems with the court were as yet unresolved, Recorde could not afford to draw attention to himself. Unfortunately it was not long before circumstances forced him to choose between his faith and his safety.

Mary's reign began rigorously with the arrest and imprisonment of great numbers of people and Recorde learnt that his friend Edward Underhill was one of those incarcerated.[2] Underhill, described by his associates as 'a witty and facetious gentleman', incautiously published a ballad mocking and attacking the Catholic church just at the moment

Mary ascended the throne. This could not be ignored and the sheriff of Middlesex, with a company of halberdiers, was sent to arrest him and bring him before the Privy Council then assembled in the Tower. Underhill, the Protestant hot-gospeller, found himself arraigned before representatives of his many Catholic enemies, who accused him of being the greatest heretic in London and forthwith committed him to Newgate Prison. After only two weeks in the gaol, his constitution perhaps weakened by the stress of heresy charges and the awful punishments meted out to offenders, Underhill became seriously ill. Recorde immediately hastened to his side and diagnosed a burning ague, a term then generally used to describe a life-threatening fever, frequently brought on by drinking contaminated water. Recorde cared for his friend with generous and manifest devotion, surreptitiously slipping in and out of the prison as occasion required.

During Underhill's sickness his wife Joan petitioned the Privy Council for his release, which was granted provided sufficient sureties were given that he would appear when called to answer the charges against him. He was carried from the prison on a horse-litter, in such a weakened state that he could hardly bear the movement of the horses, and his wife lamented 'that he could not live till he came home'. But live he did, and Underhill attributed his recovery to Recorde, saying that he was 'a doctor in physic and very learned, who ventured several times to visit me in prison to his great peril, if it had been known, who was at charges and pains with me gratis. By whose means, and God's providence, I recovered my health'. Given the extensive spy networks established under all the Tudor monarchs, it would have been remarkable if Recorde's visits to Newgate really were unobserved and unreported.

Recorde's support for Protestantism could not have been an absolute secret, but now that he had openly visited and helped a 'heretic-knave', as Underhill's enemies called him, it probably compounded his problems with the court. By the beginning of 1554 Mary had re-established the Catholic Church and promulgated new laws against heresy. The Mass, holy days, celibate clergy and Roman Catholic bishops were restored and in November of that year Cardinal Reginald Pole, Mary's newly appointed Archbishop of Canterbury, arrived at Westminster. Recorde had to do something to bolster his standing and show his loyalty to the Crown and he had to do it quickly.

He had laid aside and not yet completed the *Castle*, his book on astronomy. His original intention had probably been to dedicate the treatise to the Muscovy Company, but anxiety and agitation now caused him to change his mind. He realised that the book provided a perfect opportunity to declare publicly his support for the new regime by dedicating it to Queen Mary. Addressing her as 'most dread sovereign Lady', he was careful to distance himself from the Protestant faction that had attempted to thwart her accession to the throne, telling her that he:

> thought it my duty to make most humble suit unto your excellent Majesty... Unto which suit I am the more emboldened, through remembrance how God in despite of cankered malice and of frowning fortune, did exalt your Majesty to that throne royal, which of justice did belong unto your Highness, although the muses of mischief wrought much to the contrary.[3]

Recorde declared in the last paragraph of the dedication that he was 'animated' to finish the book and publish it under Mary's name, if she would accept it. He also cunningly signified his acceptance of her Catholic reforms by wishing her long to reign amongst them. He concluded by naming himself as 'your Majesty's most humble subject, Robert Recorde Physician':

> It may therefore please your Majesty, for love unto Knowledge, and favour to your Highness' subjects, to accept this simple *Castle* into your grace's defence, and so shall I be animated to finish the rest, and to publish it under your Majesty's name, whom God of his mercy increase in all honour royal, and true felicity, and continue prosperously and long amongst us. Amen.

Apprehensive that he had still not done enough to establish himself as a 'most humble subject', Recorde added a salutation to Cardinal Pole immediately following the dedication to Mary. He flattered the archbishop by writing in Latin with Greek interpolations, as one scholar to another, addressing him as Cardinali Polo, Cantvariensi Archiepiscopo, and Latinising himself as Robertus Recordus, Medicus.

Recorde's premonition that Protestants would be singled out and hounded was soon realised and by the summer of 1555 the Marian persecutions were in full swing. Pole's predecessor, Archbishop Cranmer, had been tried and condemned for heresy and in March the following year he died a martyr's death at the stake in Oxford. Bishops Hugh Latimer and Nicholas Ridley were also burnt alive in Oxford and throughout the country some three hundred Protestants, from all walks of life, were consigned to the flames. Recorde's fears for his safety were far from groundless.

The knowledge that his old adversary Sir William Herbert was gaining favour with the queen probably disturbed Recorde as much as it vexed him. Herbert had long been a courtier as well as a ruthless soldier, often useful and much valued by previous Tudor monarchs. Married to Anne Parr, his career had advanced dramatically when Henry VIII took Anne's sister Katherine as his sixth and last wife. He profited materially from Henry's dissolution of the monasteries and furthered the reformation by leading troops into Wales to destroy the famous shrine of the Virgin at Penrhys, in the Rhondda valley. The accumulation of offices and lands made Herbert a powerful man in Wales. In 1551 he was made lord lieutenant for all the Welsh counties and on 10 October of that year was ennobled as Baron Herbert of Cardiff. On the very next day he was elevated again and became the earl of Pembroke (hereafter simply called Pembroke).

In 1552 Pembroke had supported the plot to divert the succession from Princess Mary to Lady Jane Grey for as long as he thought it had any chance of success. He was a signatory of a letter to Mary ordering her to cease resistance and he had Jane proclaimed queen in Wales, at Beaumaris and at Denbigh. He even stood godfather, with Jane and her father, the duke of Suffolk, to the daughter of that most ardent of Protestants and Recorde's friend, Edward Underhill. However, as support for Mary grew, Pembroke began to have second thoughts,

FIGURE 17 Sir William Herbert, first Earl of Pembroke

Robert Recorde's nemesis, the nobleman, soldier and courtier who was said to have intense ambition coupled with an equally fierce temper and a hot-headed nature. According to John Aubrey's *Brief Lives*, he was 'of good natural parts, but very choleric', and was a 'mad fighting fellow,' who could 'neither read nor write' but 'had a stamp for his name'. Aubrey also reported that he had 'a little cur-dog which loved him, and the earl loved the dog'.

and together with the earl of Arundel he formed a party of defectors. Members met at Pembroke's home and there, after Arundel had given a lengthy speech setting out Mary's claim to the throne, he gave a much shorter one declaring to all present that 'if my Lord of Arundel's persuasions cannot prevail with you, either this sword shall make Mary queen, or he lose me my life'.

Mary was no fool, however, and she took more account of Pembroke's slowness in coming to her aid than of his eventual rallying to her cause. She placed him under house arrest, but he was very soon able to redeem himself. His opportunity came when Thomas Wyatt, a minor nobleman, opposed Mary's decision to marry Philip of Spain. With twenty thousand men backing him, he began a march on London in what became known as the Wyatt Rebellion. Notwithstanding the cowardice and treachery of many of Mary's sworn defenders, Pembroke kept his nerve and blocked the approaches to the city, confident that Wyatt's men would lose heart and melt away, which they did. After this episode Pembroke's experience, power and wealth probably made him indispensible to Mary and her government.

Recorde's resentment against Pembroke, a man he considered corrupt and the main source of his misfortunes, could have done nothing to ease his state of mind. It must have been galling for him to see Pembroke rise to such pre-eminence at court, whilst all his own endeavours on behalf of the Crown had brought him nothing but odium and misery. Although determined to take some action against Pembroke and make him answer for his dubious and less than honest dealings, it was not clear how he could attack such a powerful man. He seems not to have considered for a moment the wisdom of such an action, but for the present, he put the matter aside.

Doubtless with great relief, he turned to two of his old and pleasing preoccupations, mathematics and the education of his fellow men. Once more inking his pen, he commenced a book that would not only be a continuation of his first treatise on arithmetic, but would go on to introduce the esoteric subject of algebra for the first time to the English reading public. As he wrote in the preface, 'that as I have done in other arts, so in this I am the first venturer, in these dark matters'.

It has been repeated, almost *ad infinitum*, that Recorde's title for his new book, *The Whetstone of Witte*, was a clever pun. According to the accepted version of the tale, *cosa* is supposedly a Latin word meaning 'thing'. Early German algebraists are said to have used the term to label an unknown quantity, first giving it a German spelling as *coss*. It is certainly true that for many years algebraists were known as cossists because of this usage. Now, so the story goes, *coss* sounds like the word *cos*, the Latin word for a grindstone or whetstone used for sharpening knives and edge tools. Hence Recorde's supposed pun; his treatise on algebra – the *cossic* art – was a book on which to sharpen one's mathematical wit. Sadly, this ingenious play on words is not due to Recorde but originated with the mathematician Augustus De Morgan who, writing in 1847, some three hundred years after Recorde's death, noticed the similarities between the words and suggested the pun.

Because of the persistence of this myth, it is worth noting here that *cosa* is not Latin, but is an Italian word, still in use today, meaning 'thing'. When Recorde wrote 'Of the rule of Cose', he was using the plural form of *cosa* (*cose*), meaning 'things'. By adopting the Italian word to serve his own ends, we might speculate that he was acknowledging the pioneering efforts of the Italian abacists which culminated with the work of Luca Pacioli. He further underlines his recognition of algebra's history by titling a chapter in the *Whetstone* as 'The rule of equation, commonly called Algeber's Rule', thus pushing the beginnings even further back in time to Al-Khwarizmi and other Islamic mathematicians. Recorde, of course, possessed the classical education and great learning, combined with deep mathematical knowledge, that were the necessary prerequisites to including in his book these subtle pointers to the origins of his subject.[4]

Dismissing then the often repeated claim that a pun was intended, we might ask how he came to choose *The Whetstone of Witte* as the title for his book on algebra, since it is not an obvious one. He probably did not coin the phrase especially for this purpose, because in the much earlier *Pathway* he writes in the epistle to the king of 'witty men', who 'I trust by the reading of witty arts (which be as the whetstones of wit) they must needs increase more and more in wisdom'. There is no

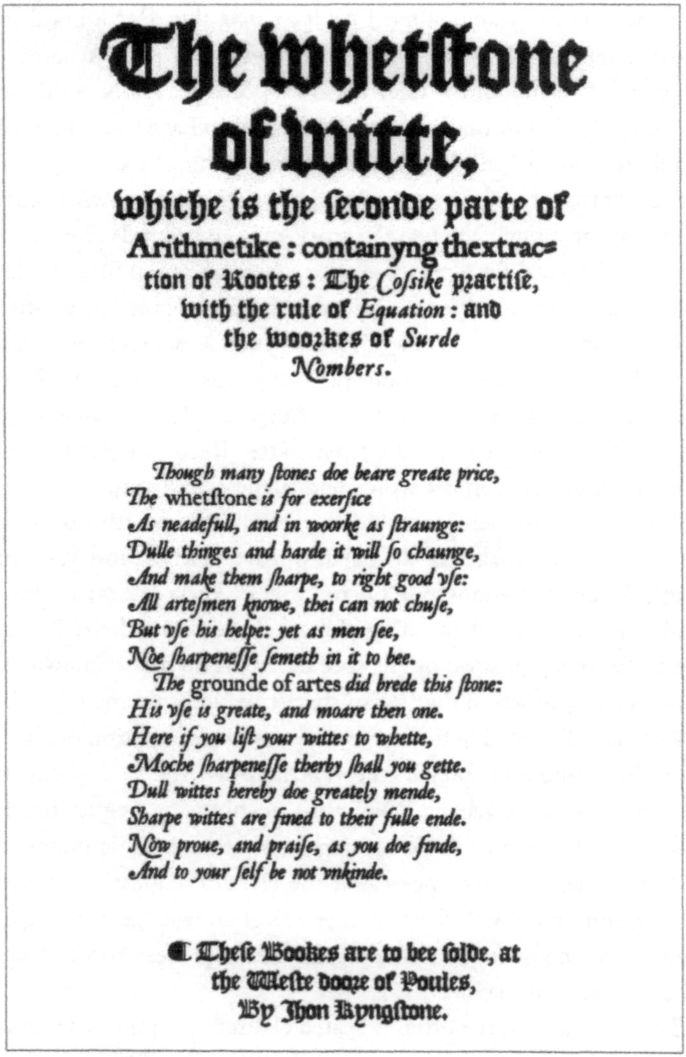

> # The whetstone of witte,
>
> ### whiche is the seconde parte of
> Arithmetike : containyng thextrac-
> tion of Rootes : The *Cossike* practise,
> with the rule of *Equation*: and
> the woorkes of *Surde*
> *Nombers.*
>
> *Though many stones doe beare greate price,*
> *The whetstone is for exersice*
> *As neadefull, and in woorke as straunge:*
> *Dulle thinges and harde it will so chaunge,*
> *And make them sharpe, to right good vse:*
> *All artesmen knowe, thei can not chuse,*
> *But vse his helpe: yet as men see,*
> *Noe sharpenesse semeth in it to bee.*
> *The grounde of artes did brede this stone:*
> *His vse is greate, and moare then one.*
> *Here if you list your wittes to whette,*
> *Moche sharpenesse therby shall you gette.*
> *Dull wittes hereby doe greately mende,*
> *Sharpe wittes are fined to their fulle ende.*
> *Now proue, and praise, as you doe finde,*
> *And to your self be not vnkinde.*
>
> ¶ These Bookes are to bee solde, at
> the Weste doore of Poules,
> By Jhon Kyngstone.

FIGURE 18 Recorde's second part of arithmetic

A continuation of his first book, *The Grounde of Arts*, this was the first book in English to deal with 'the cossic practice', and 'the rule of equation', or in modern language, algebra. Surd numbers, an expression still used by mathematicians today, are those numbers which have an infinite number of decimal places. (*The Whetstone of Witte* (1557), title page.)

mention here of algebra, German *coss* or Latin *cos*, so it would seem that when contemplating a title for his book Recorde seized upon a likely and pre-existing phrase from his earlier writings, without any thought or intention of punning.

At the beginning of the *Whetstone* Recorde once more engages in poetry, to suggest the marvellous ability of algebra to reveal something previously unknown from that which is already known. He titled the poem *Of the rule of Cose*:

> One thing is nothing, the proverb is,
> Which in some cases does not miss.
> Yet here by working with one thing,
> Such knowledge does from one root spring,
> That one thing may with right good skill,
> Compare with all thing. And you will
> The practice learn, you shall soon see,
> What things by one thing known may be.

In a following rhyme, which he called *The curious scanner*, he suggested that if any faults were in the book, readers should take pains themselves to correct them, instead of idly blaming him:

> If you ought find, as some men may,
> That you can mend, I shall you pray,
> To take some pain, so grace may send,
> This work to grow to perfect end.
>
> But if you mend not that you blame,
> I win the praise, and you the shame.
> Therefore be wise, and learn before,
> Since slander hurts itself most sore.

Like Recorde's earlier works, the *Whetstone* is written in dialogue form, the two interlocutors as before being the master and the scholar. On the first page Recorde has the scholar observe that:

the ignorant ... envy that knowledge which they cannot attain, and wish all men ignorant like unto themselves, but all gentle natures condemn such malice, and despise them as blind worms, whom nature does plague, to stay the poison of their venomous sting.

Was Recorde sniping here at his nemesis, the earl of Pembroke, who was reputed to find reading and writing difficult?

15

A HEART SO OPPRESSED

Algebra is considered to be an analytical science, and it is not therefore inappropriate to nominate Recorde, author of the first English work on algebra, as the founder of analytical science in England. According to James Halliwell, *The Whetstone of Witte* is more than a mere elementary compilation, being a work that ranks with the ablest German and Italian contemporary productions on algebra for originality and depth.[1] Furthermore, it is a technical treatise redolent of a highly trained mathematical mind. Yet this book, ostensibly concerned only with symbols, numbers and calculation, also contains amazing indications of the author's state of mind whilst writing it. For instance, describing to the scholar the mathematical concept and uses of proportions, the master suddenly interrupts himself with an observation:

> MASTER: But as for the works of proportions, I will omit them until another time, considering not only the troublesome condition of my unquiet estate, but also the convenient order of teaching. Whereby it is required that the extraction of roots should go orderly before the art of proportion, which without those other cannot be wrought.

With hindsight we know that Recorde was a deeply troubled man while writing the *Whetstone*, but the reference to the 'troublesome condition' of his 'unquiet estate' must have left his contemporary readers scratching their heads.

A few pages on the master explains the mathematical concept of the extraction of roots and then provides the scholar with a table of comparative weights of various materials, preparatory to setting him a problem to be solved.

> MASTER: And now for the use of this table, take this question. I would have five weights of cubic form, made of these five stuffs. The weight of wood shall be 28 pounds, the stone 56 pounds, the iron 112 pounds, the brass 224 pounds and the lead 448 pounds. Of all these I have but the iron weight whose side, or cubic root, is 12⅔ inches. And my desire is to know, of what quantity the sides of all the other weights shall be.
>
> SCHOLAR: The question is pleasant, and yet somewhat harder than the others.

What followed must have seemed inexplicable to readers as they struggled to follow the mathematical reasoning expounded by Recorde. Because of troubles unspecified, the master says that the scholar must work out the answer for himself, and what is more, other than the provision of a look-up table, he can expect no further tuition in the subject.

> MASTER: The table will help you fully, so that you confer it well, with what you have learned before. But because I have little leisure to spend much time with you (save that zeal to your furtherance does make me partly to forget my own business) therefore will I leave this question to yourself, to be answered at your leisure. And so in all the rest, I must pass it over and give an eye to such matters that touch me more nigh, and weigh more heavily than all such weights by twenty-fold. Wherefore, touching all the roots of compound numbers, you shall at my hand now, have no private declaration but such as you have learned already.

In this passage, Recorde reveals the stress he was under as he endeavoured to write the *Whetstone*. His troubles weighed upon him 'twenty-fold' more than all the weight of wood, stone, iron, brass and lead put together. Repeatedly we find examples of how he was quite unable to concentrate on the task in hand, and could not prevent his thoughts and his pen from wandering. About halfway through the book Recorde provides another look-up table showing all the squares, cubes and so on, for all numbers up to forty. He labels it 'The fruitful table, which may be called the table of ease'. He then has the master address the scholar in the following manner:

> MASTER: And if you like to enlarge this table, you may easily do it, multiplying the numbers still by their roots, which be set over them, in the head of the table. And so may you make it to extend infinitely, which shall ease you wonderfully, in the extraction of any kind of roots. For which at some other time if my leisure serve me better with quietness, I will give you more special rules. And also I counsel you well to examine this table, and trust not to my casting. For haste and other troubles, may often times cause error in supputation.

Supputation is an archaic word meaning 'calculation', but more extraordinary is Recorde's exhortation to his readers to distrust the figures in his table because he cannot guarantee their accuracy. This is surely a grave admission for the author of a mathematical textbook and a further indication of the great strain he was under whilst writing it.

Added to his mental agitation was the difficulty of explaining the concepts of algebra for the first time in the English language, not least because the simple mathematical notation we use today had not yet been devised. Most people will remember from their schooldays that in algebra any number can be represented by the symbol x. To multiply two such numbers together, that is $x \times x$, we use the mathematical shorthand x^2. In the same manner, to multiply three unknowns together, as in $x \times x \times x$, we simply write x^3, with $x \times x \times x \times x$ being written x^4 and so on. It is usual to read x^2 as 'x squared' since that formula gives us the area of a square. Likewise, x^3 is usually read as 'x cubed' because the formula gives us the volume of a cube. The small superscript number which indicates how many times x should be multiplied by itself we call a 'power'. This allows the discussion of numbers greater than simple squares and cubes.

So when we write x^4, we read it as x raised to the power of 4, when we write x^5, we read it as x raised to the power of 5, and so on. For mathematicians of Recorde's time there was no such easy way of denoting the powers of numbers, which was a great hindrance to the development of effective mathematics. Yet he needed to be able to express the concept of a number raised to a power if he was to complete his book. His solution was to follow the examples of continental mathematicians, adapting their methods to his own use and to the English language.

In 1494 the Franciscan friar Luca Pacioli, a teacher of mathematics to Leonardo da Vinci, published his *Summa de arithmetica, geometria, proportioni et proportionalita*. This was a synthesis of all the mathematical knowledge of his time, and it contained the first printed treatise on algebra. Pacioli used the Italian word *cento* to indicate the square of an unknown number – x^2 in modern terminology. German algebraists seized on Pacioli's usage, first changing the letter 'c' to 'z' the better to imitate the Italian sound of the word, and then Latinising the resulting *zento* to *zensus*. Recorde adopted this word, but because he was writing in English and not Latin he spelt it 'zenzike', telling the scholar that 'you shall understand that many men do ever more call square numbers zenzikes'.[2] To express a number raised to the power of 4 he wrote, by extension, zenzizenzike, and to the power of 8 zenzizenzizenzike. Numbers raised to the power of 6 he wrote as zenzicubic and to the power of 12 as zenzizenzicubic. In the 'fruitful table' he even wrote zenzizenzizenzizenzike to expresses a number raised to the power of 16, although in order to fit this long word into the available space the printer abbreviated it by omitting each letter 'n' and placing a diacritical mark over the preceding letter 'e' to signify the omission, thus – zēzizēzizēzizēzike. Thankfully Recorde pushed this terminology no further, but the bewilderment of his readers on encountering it for the first time can be imagined (some mathematical explanations may well cause a similar bafflement in today's students). Now long obsolete, zenzizenzizenzike (x^8) survives in the *Oxford English Dictionary* as an historical curiosity with a single citation – Recorde's book. It turns up from time to time as a quiz question, allegedly containing the letter 'z' more times than any other word in the English language.

FIGURE 19 Recorde's table of algebraic symbols

A listing of the complicated notation for numbers raised to a power. The symbol under the number 2, for example, equates to the modern notation x^2. Likewise, that under 3 equates to x^3 and that under 49 would today be written simply as x^{49}. Recorde's readers may have been thankful that he stopped at x^{80}. (*The Whetstone of Witte* (1557), sig. V.i.r.)

There is little doubt that Recorde struggled with mathematical notation in the *Whetstone*. Without our modern symbolism of x raised to a power, he was obliged to use peculiar tokens in his calculations, somewhat resembling Greek or Coptic letters, to represent every single instance of a power. So, for example, there was a token or symbol for x, a different one for x^2, another for x^3 and so on, up to x^{88}, which Recorde decided was quite far enough. No doubt his readers were greatly relieved, since they had to learn how to combine the basic symbols for squares, cubes and powers of five in order to form the symbols for all the higher powers before they could make further progress in their studies. He also adopted symbols for plus and minus from German writers, telling the scholar that 'there be another two signs in often use, of which the first is made thus + and betokens more, the other is thus made − and betokens less'. This was the first usage of our familiar plus and minus signs in an English language book. Recorde portrayed the scholar as quite brilliant at mastering the algebraic symbols, on the principle that if he could do it, so could any other learner.

Recorde did more than just use mathematical notation devised by others. R. T. Jenkins, writing in the 1920s, said that every schoolboy knew Robert Recorde's name as that of the inventor of the 'equals' sign, the mathematical shorthand for which is the familiar =. However well known Recorde's connection with the sign was then, it is probably even better known today, especially in Wales, where schoolchildren have adopted Recorde as a 'Welsh hero' because of this omnipresent symbol which he devised. Seeking to instruct the scholar in the art of solving simple equations, where it is necessary to set one expression equal to another, Recorde began by having the master explain the rationale of his invention.

> MASTER: And to avoid the tedious repetition of these words 'is equal to' I will set as I do often in work use, a pair of parallels, or gemowe lines of one length, thus =, because no 2 things can be more equal.

Recorde adapted the word 'gemowe' from the Latin *gemellus*, meaning twin lines, a word having the same origin as Gemini, the twins, in the signs of the zodiac. In books printed before Recorde wrote the *Whetstone*, equality was often expressed by writing the Latin word *aequales*, sometimes abbreviated to *aeq*, as for example 2 + 2 *aeq* 4. Recorde's symbol of two parallel lines was not immediately taken up by other mathematicians and it might easily have disappeared in competition with other symbols used on the continent. It was to be over seventy-five years before two parallel lines received general recognition in England as the symbol for equality, being adopted in three influential works, Thomas Harriot's *Artis Analyticae Praxis*, William Oughtred's *Clavis Mathematicae* and Richard Norwood's *Trigonometric*. Its use spread slowly from England to Europe and, ultimately, to the rest of the world. Today it is difficult to imagine that there was once a time when Recorde's pair of parallels = was not the sign of equality.

So heavily did anxiety and worry impinge on Recorde's mind that, as he neared the end of the *Whetstone*, he was clearly perturbed and perhaps close to a breakdown. On the penultimate page, in the middle of a discussion on the difficult subject of surd numbers, he abruptly finished and penned an exchange between the scholar and the master that must surely be the most extraordinary ending to a mathematics textbook ever written:

> SCHOLAR: Now I perceive that in Addition and Subtraction of Surds, the last numbers that did result of that work were universal roots.
>
> MASTER: You say truth. But hark, what means that hasty knocking at the door?
>
> SCHOLAR: It is a messenger.
>
> MASTER: What is the message? Tell me in mine ear. Yea sir, is that the matter? Then is there no remedy, but that I must neglect all studies and teaching, for to withstand those dangers. My fortune is not so good, to have quiet time to teach.
>
> SCHOLAR: But my fortune and my fellows is much worse, that your unquietness so hinders our knowledge. I pray God amend it.

> ### The Arte
>
> as their workes doe extende) to distincte it onely into twoo partes. Whereof the firste is, *when one number is equalle vnto one other.* And the seconde is *when one number is compared as equalle vnto 2. other nombers.*
>
> Alwaies willyng you to remeber, that you reduce your nombers, to their leaste denominations, and smalleste formes, before you procede any farther.
>
> And again, if your *equation* be soche, that the greateste denomination *Cossike*, be ioined to any parte of a compounde nomber, you shall tourne it so, that the nomber of the greateste signe alone, maie stande as equalle to the reste.
>
> And this is all that neadeth to be taughte, concernyng this woorke.
>
> Howbeit, for easie alteratiō of *equations*. I will propounde a fewe exāples, bicause the extraction of their rootes, maie the more aptly bee wroughte. And to auoide the tediouse repetition of these woordes : is equalle to : I will sette as I doe often in woorke vse, a paire of paralleles, or Gemowe lines of one lengthe, thus : ══════ bicause noe. 2. thynges, can be moare equalle. And now marke these nombers.
>
> 1.   14.ze. ——— .15.℥ ═════ 71.℥.
>
> 2.   20.ze. ——— .18.℥ ═════ .102.℥.
>
> 3.   26.℥. ——— 10ze ═════ 9.℥. ——— 10ze ——— 213.℥.
>
> 4.   19.ze ——— 192.℥ ═════ 10℥. ——— 108℥ ——— 19ze
>
> 5.   18.ze ——— 24.℥. ═════ 8.℥. ——— 2.ze.
>
> 6.   34℥. ——— 12ze ═════ 40ze ——— 480℥ ——— 9.℥
>
> 1.   In the firste there appeareth. 2. nombers, that is 14.ze.

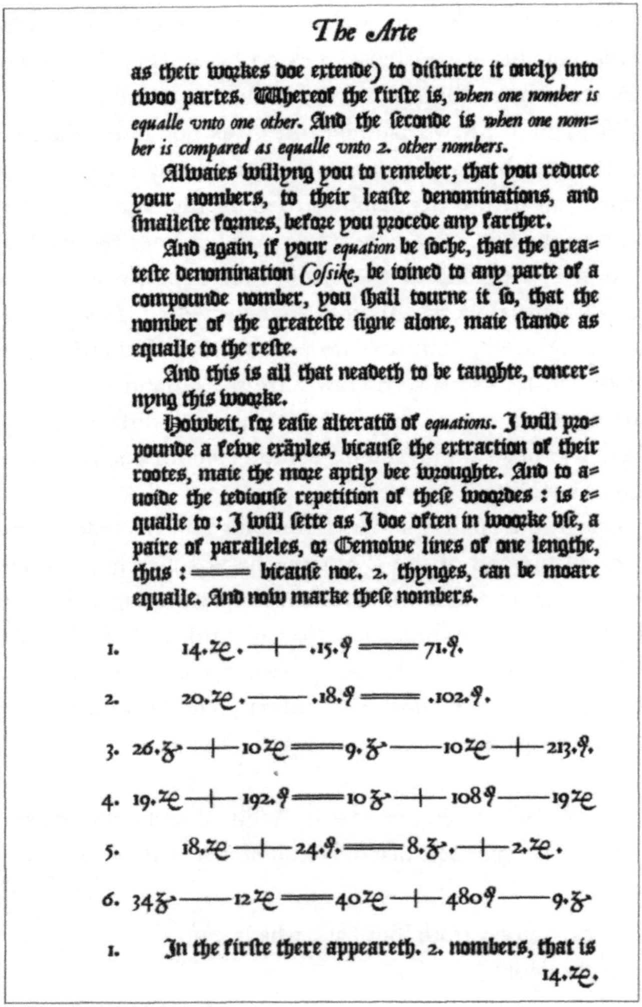

FIGURE 20 Recorde devises the sign for equality

The notable page on which Recorde explains his use of two parallel lines, writing that 'to avoid the tedious repetition of these words – is equal to – I will set as I do often in work use, a pair of parallels, or Gemowe lines of one length, thus: ══ because no two things can be more equal'. The six equations which follow show the first ever use of parallel lines as the symbol of equality in a printed book. Today, mathematicians would write equation 6, for example, in this manner: $34x^2 - 12x = 40x + 480 - 9x^2$. (*The Whetstone of Witte* (1557), sig. Ff.i, *v*.)

MASTER: I am enforced to make an end of this matter. But yet will I promise you that which you shall challenge of me, when you see me at better leisure. That I will teach you the whole art of universal roots. And the extraction of roots in all Square Surds, with the demonstration of them, and all the former works. If I might have been quietly permitted to rest but a little while longer, I had determined not to have ceased, till I had ended all these things at large. But now farewell. And apply your study diligently in this that you have learned. And if I may get any quietness reasonable, I will not forget to perform my promise with an augmentation.

SCHOLAR: My heart is so oppressed with pensiveness, by this sudden unquietness, that I cannot express my grief. But I will pray, with all them that love honest knowledge, that God of his mercy will soon end your troubles, and grant you such rest, as your travail does merit. And all that love learning, say thereto. Amen.

MASTER: Amen, and Amen.

As events will show, this farewell is a most poignant and prophetic ending to the last book Recorde wrote. He does not elaborate on the dangers he must withstand, or tell us why he is not permitted to rest a little while longer, but evidently he feared a knock at his door that would bring 'unquietness' and distress into his life if, or when, it came.

There were three possible causes for his fears. The first was the Marian persecutions, and the experience of Richard Eden, his acquaintance and his brother's friend, showed that there was a need to be extremely careful during the dangerous times of Mary's reign. Eden, a Protestant, had translated into English an apparently innocuous book from the continent about recent geographical discoveries. He prefaced the volume with long and fulsome praise of Mary and her husband Philip of Spain, obviously with the intention of ingratiating himself with the two monarchs. Unfortunately for him, an allegedly objectionable passage in the preface caught the eye of zealous Catholic churchmen. He was promptly dismissed from a lucrative position in Philip's treasury and hauled before the Bishop of Winchester, Stephen Gardiner, on a charge of heresy. This incident, and the remembrance that Edward

Underhill had also been imprisoned on a charge of heresy, could not have been lost on Recorde. It must have seemed to him that he stood perilously close to martyrdom and this very real danger may have been the one he referred to in his address to the scholar.

The second of Recorde's worries was the impending silver mines inquiry, about which he could do nothing except patiently await developments. His third worry was his quarrel with Pembroke and, perhaps overburdened by despondency, he decided to tackle this problem head-on. On 10 June 1556 he wrote a missive to the queen complaining about Pembroke's charge of treason against him and asking that his side of the story be heard. Etiquette demanded that the queen should not be addressed directly, but through one of her councillors, and he chose William Ryce, a member of the queen's Privy Chamber with whom he was evidently acquainted. He fully expected that his letter would be brought to her attention:

> To the right worshipful Mr Ryce, one of the gentlemen of the Queen's majesty's privy chamber.
>
> Sir, I am right sorry that the malice of any man should hinder the declaration of my good will, namely where I was so willing to have expressed it, and yet such is the chance presently. For since my last being with you the Earl of Pembroke has commenced against me an action of £1,000 albeit I never had to do with him for myself but for King Edward only whom God pardon. But I think it had been better not only for his majesty but also for me if neither of us both had known the good earl. As long as I was in office and might answer him accordingly he would never attempt any action against me directly, although craftily he with his confederate of Northumberland with great prejudice to the Crown made wonderful attempts for my destruction, objecting against me first treason and then heinous contempt, when if justice had been free himself had been worthier the rewards of both. And therefore as long as I was absent from the Court he thought it not good to wake a sleeping dog, but seeing now, mere suspecting my repair thither again

seeks means to stay me some other ways, lest by the liberty of my tongue he might be as well known to other of the Queen's highness true friends (*absit verbo arrogancia*)* as he is unto me. I have written to you earnestly and am willing to answer it as gladly which redound to the Queen's majesty great commodity if the matter be handled accordingly, and namely such circumspection *usid ne mutuo stanat muli*.** But to conclude: *testes advoco celum sidera ac sapientes nunquam dissidium in Brittania extinctum iri donec occulte inimicite prosapie veri et fucati draconis palam erumpant*.*** Knowing your faithfulness to the Queen's highness with liberty of access to the same this much have I written with my own hand. *Quelicet occasio precosior sit amplectanda tamen est*.**** But to the intent it may not be thought that I seek aid against the Earl I wish I might answer him before the privy council, for the matter touches the Queen's majesty whom God prosper.

yours fully in that he can do. R. Recorde.[3]

That Recorde chose to attack Pembroke at a time when he had risen to become one of the most powerful men in England and Wales, and was held in the highest esteem by Mary, is a further indication of the effect his troubles had on his usual clear thinking. He exhibited a complete lack of that judgement he had so often exercised as a physician and as a renowned academic and scholar. Moreover, he was about to attract Mary's attention to himself, and she could only have regarded him as an intellectual Protestant heretic, worthy of repression for that reason alone. It was certainly a bad move on his part, but the letter was sent, and he could only await the outcome with trepidation, probably to the further detriment of his nervous health.

---

* speaking without arrogance.
** is not used to prevent mutual support.
*** summoning as witnesses the heavens, the stars and wise men, never will bitter dissent be extinguished in Britain as long as there is hidden hostility to the true lineage, and a painted dragon may openly emerge.
**** Although it is a precious opportunity to be embraced.

16

AN UNQUIET MIND

Prospects must have seemed very bleak in Recorde's 'unquiet mind' at the start of 1556, but then two events occurred which surely gave him some cheer. First, the senate of Cambridge University, unconcerned by his long absence from the campus, honoured him with the ceremony of Vestro Communi. During the proceedings he was officially recognised by the university as a member of the distinguished community of Cambridge scholars. We do not know whether he attended the ceremony in person, or learnt of it by letter, but either way it surely lifted his spirits to know that his reputation was undamaged and stood as high as ever among his academic peers.

Secondly, he had finally finished writing *The Castle of Knowledge*. Recorde included more than eighty illustrations and diagrams in the book, some of them of a complicated nature, and it may be that completion was delayed due to the engraving of plates as much as by his tardiness in finishing the text. We can surmise that this book took Recorde a long time to complete from the short poem he placed on the contents page:

> If ought here want, that you desire,
> Remember where this work was wrought.
> In Pluto's forge with scarce good fire,
> This rusty Sphere to end was brought.
> But if I may it file again,
> The rust I trust to scour off clean.

The sphere referred to was, of course, the celestial sphere, the subject of the book. That it had become rusty and needed metaphorical scouring and cleaning indicates a long period of inactivity and neglect. The exhortation to remember where the book was written (wrought) is more cryptic. Recorde's allusion to Pluto's forge seems to be hinting that he completed the book under difficult circumstances, since the operation of a forge without a good fire would be very difficult indeed. If this interpretation is correct, the difficulties under which he laboured may refer to the troubles that were beginning to overshadow his life.

Perhaps because of its long gestation, Recorde might have forgotten that at the end of the book he had railed against 'infidels and false Christians'.[1] It seems not to have been noticed by the ardent Catholic churchmen around Mary, or they would surely have made much of it, coming from the pen of a Protestant writer in a book dedicated to her. It was so easy to make a false step, leading to a speedy downfall, as Richard Eden had discovered to his great cost. However, Reyner Wolfe finally published the *Castle* around September 1556 and it proved to be a great success.

In the latter half of that year Recorde inexplicably added to his problems by insulting the earl of Pembroke with a charge of malfeasance. He may have been motivated to take this reckless action by the hope that Pembroke would drop his accusations of treason against him, in return for the withdrawal of his own imputations regarding the earl's alleged wrongdoings. If so, he was mistaken and his counter-accusations, although probably valid, had no hope of success, not least because of Pembroke's closeness to the monarch. In the book's dedication to the queen, Recorde pleaded that:

> although I could not be permitted by disturbance of cruel fortune, to accomplish now my building [writing the *Castle*] as I had drawn the plan, yet in despite of fortune this much have I done, which is more than ever was done in this tongue [English] before, as far as I can hear.

It seems reasonable to connect the 'disturbance of cruel fortune' with the recognisance, or summons, that he received about six weeks after

192   THE FOVRTHE TREATISE OF

Howe be it bicaufe you fhall know what names thelder grekes dyd giue them (whyche names hath beene retayned euer fith that time) I haue here drawen a lyke table as your other authors haue fette forthe, that you may the better conferre the figure with the table, and the more eafilye vnderftande the one by the other. in whiche figure the cirde A, B, C, D, represeteth the Horizont, & the righte line A C, ftandeth for the Meridiane line. A is y̆ north pole and C, the fouth pole. B the eafte, & D y̆ weft. B D betokening the Equinoctiall, and E F the tropike of Cancer, GH, the tropike of Capricorne. and al the other lines are the boundes of the Climats eche in his order. The firft Climat taketh name of Meroe, a famous Iland in Ethiopia vnder Egypt, indofed by the riuer Nilus. the fecōd Climat is named of Syene, a city of Egypt, lying directli vnder y̆ tropik of Cancer. The third Climate is called after Alexādria, a notable city & an anciēt vniuerfity in egypt alfo, lying on the north fhore of it. The fourth dimate beareth y̆ name of y̆ Rodes, an iffand better knowē then kept, and yet better lofte then kepte fo derely. The fifte Climate is expreffed by the name of Rome, a citye in Italye well ynoughe knowen.
                                                                                           The

FIGURE 21 Recorde's discussion of 'climates', or time zones

On this page Recorde shows his knowledge of European geography and contemporary events. He names Ethiopia and Egypt, and the cities of Meroe, Syene (Aswan) and Alexandria. He also mentions Rhodes, 'an island better known than kept, and yet better lost than kept so derely'. This casual reference is to an event in 1522, when Europe was shocked by the sacking and loss of Rhodes to the Saracens, after the defending Knights Hospitallers suffered great deprivation and dreadful loss of life. (*The Castle of Knowledge* (1556), p. 192.)

sending his letter to Mary, ordering him when called to appear before the Privy Council. The queen had read his letter and passed it on to her privy councillors, who issued the recognisance from St James' Palace, dated 20 July 1556, under the signatures of the lord chancellor, the lord treasurer, the bishop of Ely, Mr Secretary Petrie, Mr Secretary Bourne, Sir Francis Inglefield, Sir John Baker and, significantly, the earl of Pembroke. A stipulation of the recognisance was that Recorde should nominate two citizens to appear with him, each prepared to stand bond in the sum of £1,000 against his fleeing before being called. Reyner Wolfe came forward with a surety for his friend's appearance, as did a certain John Wykes, a goldsmith and citizen of London. Nothing more is known of Wykes, but he must have been more than an acquaintance to hazard such a large sum of money on Recorde's behalf. The document stated *inter alia* that:

> The condition of this recognizance is such that if Robert Recorde, Doctor of Physick, dwelling in the parish of St Katherine Coleman in London, at all times hereafter within the space of one year next after the date hereof, upon ten days warning to be given him, or either of them [his sureties], make his personal appearance before the Lords of the King's and Queen's Majesties Privy Council to answer such matter as shall be objected unto him.[2]

We do not know what matters the Privy Council intended to place before Recorde at the hearing, but the queen's rapid response to his letter points to their seriousness. There can be little doubt that the intention was to probe the Irish mines affair and the allegations made against him, and his counter-allegations against others regarding pecuniary losses to the Crown. However, the inclusion of the bishop of Ely as one of the investigators suggests that Recorde had provoked the queen's displeasure as much for possible heresy as for mismanagement of financial matters. However, it is unlikely we shall ever discover the true intentions of the queen or the Privy Council as the hearing was never held. It was overtaken by another contingency, ostensibly a private matter between Recorde and Pembroke, and in the event neither of Recorde's sureties was called upon to honour their bonds.

On 16 October 1556 Pembroke sued Recorde for libel. Pembroke's attorney served the bill, which alleged that Recorde's letter to the queen made the earl appear a traitor, injured the Crown and expressed the opinion that he deserved imprisonment. Pembroke was certainly baffled by the Latin phrases in the letter which, given his limited scholastic abilities, he was unable to read. Nevertheless, when the lines were translated for him he may well have concluded that *fucati draconis*, the 'painted dragon', meant the red dragon of Wales, that is, himself. Since Recorde appeared to be insinuating that the dragon had a 'hidden hostility to the true linage', in other words to Mary's claim to the throne, it is not surprising that he thought Recorde was calling him a traitor. The threat of prison contained in the writ suggests Recorde's incarceration might have been the primary objective all along, and the suit for libel merely a feint to secure his person until graver charges could be instigated. There is precedent for such chicanery during Queen Mary's reign. For example, John Hooper, a former monk and a Protestant reformer, was arrested and sent to the Fleet prison for alleged debt in January 1555. He never regained his freedom and after a staged show trial which condemned him as a heretic, he suffered a particularly harrowing death by burning in Gloucester.

The lawsuit was heard at Westminster Hall in January 1557 and Recorde attempted to defend himself with a detailed list of allegations against Pembroke. He claimed that in 1549, at the Pentyrch iron workings, Pembroke had stolen Crown property and money and, by allowing his retainers to drive off the workers, had lost King Edward £2,000. Recorde also asserted that on several occasions whilst he was the comptroller at the Bristol mint, Pembroke had sent his servant Roger Wygmore to carry away £1,826 in total, purely for the earl's own use. By plotting Recorde's removal from the mint and his enforced stay at court, the earl caused the moneyers to stop work for sixty days, which cost the king £10,000. Recorde further claimed that the earl had approved the accounts of the Dublin mint without reference to himself, thereby losing the Crown £40,000, presumably through false accounting. Finally, Recorde made a general accusation that as overall supervisor of the mints Pembroke defrauded the king, through the accounts of officials, to the sum of £50,000. It was all to no avail. A nobleman as powerful as

Pembroke, indispensible to the monarch, commander of her armies, able to dispense grace and favour to all who clustered around him at court, had only to deny everything for all Recorde's charges to be dismissed. On 10 February he was adjudged guilty of libel and ordered to pay the earl £1,000 damages with costs. It was then up to Pembroke to press for payment of the fine, and if Recorde either could not or would not pay the requisite sum, then it was open to Pembroke to return to court and ask for his committal to prison as a debtor.

Recorde returned home and continued to work on the *Whetstone*. His dedication of the book to the 'Right Worshipful, the Governors, Consul's, and the rest of the company of venturers into Moscovia', is strange, since the book had no obvious connection with navigation. He may have felt that he was fulfilling an obligation, incurred when he hurriedly dedicated the *Castle* not to the worshipful governors but to Queen Mary instead. He concluded his address to the company with a stinging execration:

> God prosper well your endeavour and send you such good success, as so worthy adventure does deserve. Which I doubt not will ensue if cankered malice of some spiteful stomachs do not prevail, as they cannot cease to practice to hinder your commodity, and deface your travail. But as it is ever seen, and therefore commonly known, that envy does still repine at glory, so ought all honest hearts to prosecute their good attempts, and condemn the bawling of dogged curs. So fare you well. And love him again, that delights and studies to further your commodity.
>
> At London, the 12 day of November 1557

It is quite probable that the bitterness evident in this paragraph is directed at Pembroke. The last sentence is particularly poignant in view of Recorde's undoubted despair as he awaited the outcome of events. However, as the dating of the dedication makes clear, he was still at liberty some ten months after the libel judgement.

The *Whetstone of Witte* was published late in 1557 by John Kyngston, the printer for whom Recorde had edited Fabyan's *Chronicles*. It is not known why Reyner Wolfe did not print it, as he had done all Recorde's other books. He may have thought it prudent to distance himself from his friend at a time when he was in great trouble with the authorities. Wolfe was, after all, a Protestant himself, and a known associate of the detested Archbishop Cranmer, who had effected the dissolution of the marriage between Henry VIII and Catherine of Aragon and thus made their daughter Mary illegitimate. In those arduous and difficult times, loyalty to friends had sometimes to be sacrificed for the sake of self-preservation.

Late in 1557, or early in 1558, the rap on the door that Recorde had so long dreaded finally came. It was the custom for the sheriff of London to be held responsible for the arrest of all offenders and it would have been his officers who applied the 'hasty knocking' that Recorde feared. Edward Underhill described how, at the time of his arrest, two of the sheriff's officers were appointed to convey him to Newgate, but 'to go a pretty distance behind him, without halberds in their hands, that the less notice might be taken of him'. Perhaps Recorde was accorded the same courtesy, though the normal procedure was for an officer to walk each side of the prisoner holding him by the arms, 'lest he should shift from them amongst the people'. Recorde was taken to the King's Bench prison at Southwark on the south bank of the Thames, almost opposite his home.

Who ordered Recorde's arrest, and for what reason, is uncertain. There is a long-held assumption, tediously repeated, that he was imprisoned for debt. This is unlikely, and there is no evidence that Pembroke ever pursued him for payment of his damages. Even supposing that he had, and Recorde had been embarrassed by a lack of ready money, he had property in Tenby that could have been sold or mortgaged. Alternatively, he had wealthy friends such as Wolfe, Whalley, Underhill and Lougher, to name only a few, from whom he could have easily raised a loan to pay off the damages on the security of £1,000 still owed to him by the government for his services in Ireland. It has been suggested that he could in fact have afforded to pay, but chose not to enrich his hated enemy. This is an anachronistic notion, based on the situation in modern times when an individual may, as a matter of principle, prefer a term in prison to paying unfair taxes and the like. In Tudor times prisons were unhealthy, debilitating and noisome, and anyone who could avoid a committal did so by any means possible. A spell of more than a few months' confinement was often tantamount to a death sentence because of the diseases then endemic in all gaols.

Since there appears to be no valid reason for the belief that Recorde was imprisoned for personal debt, nor indeed that he had been found guilty of anything warranting imprisonment, we must look to the judicial customs prevailing in Tudor times for the cause of his captivity. Reasons for committal were generally of three kinds, punitive, coercive and custodial. A punitive sentence was imposed as a punishment for criminal acts, a deprivation of liberty for a specified period of time. Coercive imprisonment usually meant a sentence without limit, typically in the case of debtors until their debts were repaid or until they became enfeebled and eventually died. Both punitive and coercive imprisonment could be imposed only by due process of law and a court sentence. Custodial imprisonment, on the other hand, could be imposed without trial. Anyone sufficiently powerful who wanted to detain a suspect in custody for an unspecified length of time whilst evidence was gathered – or fabricated – could order it. It seems likely therefore that Recorde was sent to prison on this third count. A member of the Privy Council, an officer of the government, or anyone who could claim to be acting in the queen's name could have had him gaoled without any formal action at law. It was not until 1679 that the Act of Habeas Corpus (literally 'you may have the body') became law. This writ prevented abuses of power by allowing prisoners or their friends to demand an appearance before a court to determine whether the gaoler had lawful authority to impose the detention. If the custodian was found to be acting beyond the law, then the prisoner had to be released. Recorde had no such recourse, hence perhaps his rather plaintive utterance that 'he must go learn some law'.

It is reasonable to assume, therefore, that Recorde was held in the King's Bench prison, without trial while enquiries into the Irish mines affair got underway. Queen Mary would not have wanted him harried until those matters were concluded, but she might well have required him to be held at her pleasure until her bishops were able to examine him for the taint of heresy. If this supposition is correct – and no other suggests itself – he would probably not have been confined in a dungeon-like cell but in a room with a barred window and an open door, giving access to a courtyard enclosed by a high wall. He would have had to pay for his furnished lodgings, his own food and clean clothing. The authorities did allow some inmates to live a short distance outside the prison, a practice known as the 'Liberty of the Rules', but if he was indeed a prisoner of the state Recorde was unlikely to have been granted this privilege.

He would have recalled that Underhill had languished in Newgate for only two weeks before becoming seriously ill. Accordingly, he would have taken care of his own health as best he could, but after about six months the debilitating effects of constant worry, poor food and probably contaminated drinking water brought him down, just like Underhill, with the notorious 'gaol fever'. This could have been dysentery or typhus, or perhaps typhoid fever occasioned by body lice, although these illnesses were not known by such names at the time. At the beginning of June 1558 he must have realised that he was dying and called for paper, pen and ink to make his will. It is not known whether he was well enough to write it himself or whether he dictated it to a notary:

Testament of Robert Recorde

In the name of God amen. For as much as nothing is more certain to man than death, and nothing more uncertain the hour and time thereof, therefore know you me, Robert Recorde, Doctor of Physick, though sick in body yet whole of mind, thanks be to God, make my last will and testament in manner and form following.

First I commit my soul into the hands of the same almighty God, my only Maker and Redeemer, trusting by the merits of his passion to be one of his elect in glory forever. My body as received from the earth I bequeath thither again, to be buried among other Christians according to the solemn usage of the church.

> My temporal goods and chattels I order, will and dispose in manner and form following.
>
> Secondly I give to Arthur Hilton, under-marshal of the King's Bench where I now remain prisoner, 20s.
>
> Item: To his wife, another 20s.
>
> Item: To the gentlemen now prisoners with me, 30s.
>
> Item: I give another 20s. to the said Arthur Hilton, to be by him distributed among the officers according to his discretion.
>
> Item: To his wife, to be distributed among her women, 6s. 8d.
>
> Item: I give generally to the common goal of the said prison, 40s. to be equally distributed among the prisoners there.
>
> Item: I give and bequeath to mine own mother and to my father-in-law, her husband, £40.
>
> Item: I give to my servant John, £6.
>
> Item: I give unto the children of John Battyn, 40s. to be distributed at the discretion of their said father.
>
> The residue of all my goods and chattels, moveable and immoveable, real and personal, I give and bequeath unto my brother, Richard Recorde, and Robert Recorde, his son, my nephew, whom I make and ordain my full and whole executors, to the end that they, beside my funerals of the same, shall truly and faithfully pay my debts, which are to Nicholas Fulgeham, citizen and merchant tailor of London, £50, to Mr Battyn, 40s.

Recorde gave further thought to the provisions of his will during the night and was well enough the next day to return to it and add a codicil:

> Memorandum, that the said testator on the morrow next after the making of his testament aforesaid, being then of his perfect mind and memory, adding to the said testament, gave and bequeathed to Alice and Rose Recorde, daughters of the said Richard Recorde, and also unto Julian Raye, all his utensils or household stuff, to be equally divided between them. Item: he willed and devised that Nicholas Adames, then being prisoner in the King's Bench,

should have all his books concerning the laws of this realm, at the price of £4.

Witness hereunto Richard Corbett, George Marten and Richard Thymylby.

Recorde's relationship to Mr Battyn and his children, and to Julian Raye, is unknown, although he obviously knew them well.

The original will has long since disappeared. However, it was necessary for a probate court to verify the legal validity of all wills before granting the named executors legal powers to dispose of the testator's assets in the manner specified in the will. The Prerogative Court of Canterbury, a church court under the authority of the archbishop of Canterbury, was responsible for proving wills in the whole of southern England, including London, and Wales. Recorde's will was therefore sent to Canterbury where a clerk copied it into a ledger and then, in a neat legal hand, added in Latin (translated below) the standard probate clause. It is this copy which survives today in the National Archives at Kew.[3]

> The above-written testament was proved, probated and registered, together with the codicil to the same, before the Lord [the archbishop] at London, 18 June 1558, by the oath of Richard Recorde, executor named in the same will, and administration of all and singular goods, rights and credits, etc., was granted to the aforesaid executor, to good and faithfully administer the same, and to prepare a full and faithful inventory of all goods, rights and credits, reserving the power, etc., to Robert Recorde, also an executor, etc.

With his affairs thus satisfactorily in order, Recorde must have retired thankfully to his bed. His servant John was probably in attendance and also perhaps his brother and his nephew Robert. Perhaps the 'gentlemen now prisoners with me' were also gathered around the bed of the dying man.

17

ONE OF HIS ELECT IN GLORY

Robert Recorde breathed his last not among the Cambrians, but in the close confines of the King's Bench prison, probably not long after making his will and about six months after he was imprisoned. He was forty-six years old or thereabouts, the exact date of his demise being unknown. His relatively peaceful end may be contrasted with the awful possibilities had he lived. During the Marian persecutions suspected heretics were held in prison for extended periods, often mistreated to force a recantation and then consigned to the flames despite having recanted – *vide* the awful end of Archbishop Cranmer. It may be that Recorde's untimely death was actually a blessing.

We know nothing of his burial place, but a plausible scenario is that his body was carried back over London Bridge by his brother, his nephew and his servant John and then to the parish church of St Katherine Coleman on Church Row. Here, in the graveyard of the church from whose pulpit he may have preached, he may have been buried; nothing is certain, but if so, it was not to be his last resting place. The fourteenth-century stone church that Recorde knew was replaced in about 1740 by a red brick building, which was itself demolished in 1926. When the graveyard was cleared, remains were reburied in the City of London cemetery, so if Recorde lies today in an unmarked grave, it is likely to be there. A plaque recording the site of the church can be found on Church Row, now renamed St Katherine's Row. The churchyard survives as a small public garden owned by Lloyd's Register of Shipping, overlooked by the new building for Lloyds erected in 1996. The garden is largely hard-surfaced, with two mature trees, raised flowerbeds, a fountain and seating. It is possible to sit in this quiet haven, insulated from the roar of London's traffic, and ponder the fact that somewhere near here there once lived one of Tudor England's greatest scientific minds.

Queen Mary died shortly after Recorde, on 17 November 1558, being succeeded by her half-sister Elizabeth. Had Recorde lived he might well have joined his contemporary Dr John Dee in being counted among the celebrated scholars of the dazzling Elizabethan age. It was not to be, but in 1570 lobbying by Recorde's nephew and namesake Robert brought to Elizabeth's notice the fact that the government owed Recorde money for his services in Ireland and she ordered that his family be recompensed. In exchange for Robert junior relinquishing his claim to £1,054 19½d owed to his uncle by Edward VI, he was granted a twenty-one-year lease of Crown lands in Cambridgeshire, Sussex, Caernarfonshire and Pembrokeshire having rents totalling over £25 per annum. So in death Recorde did restore the family fortunes in Tenby.

His brother and his brother's descendants continued to achieve prominence in Tenby. Richard followed in his father's footsteps and became mayor in 1559. Robert junior became a bailiff in 1584 and another of Richard's sons, Erasmus Recorde, was bailiff of the town in 1610. It is often repeated that Dr Robert Recorde had four sons and five daughters. This is untrue and is simple confusion between uncle and nephew. Recorde was, of course, a lifelong bachelor and it was his nephew of the same name who was the progenitor of so many offspring. Later in the century the family acquired as their home the old Hospital of St John the Baptist, which stood near to the present St John's Hill and St John's Croft in Tenby. There is nothing now to be found of this building, although its representation survives in an etching made in 1812 by Tenby artist Charles Norris.

Recorde predeceased John Robyns, his colleague from his younger days at Oxford, by only a month or so, Robyns dying on 25 August 1558. Robyns had lived through the reigns of Henry VIII and Edward VI, holding the posts of vicar and rector in a number of parishes, and in 1543 he became chaplain to Princess Mary (the future queen). He was buried, full of honours, in the chapel of St George at Windsor, under a marble slab engraved with a fulsome eulogy, in marked contrast with Recorde's sad end and unknown grave.

Reyner Wolfe died in December 1573, leaving his St Paul's Churchyard properties, reputed to be a continuous stretch of more than 120 feet of the best bookselling frontage in England, to his wife Joan, who briefly carried on his business. She died in 1574, leaving the Brazen Serpent printing house to their eldest son, Robert. In her will she bequeathed Wolfe's unfinished chronicle to Raphael Holinshed and asked 'to be buried decently in the parish church of St Faith in London, as near as may be unto my late dear husband Reginald Wolfe, late citizen and stationer of London'.

Richard Whalley, the dedicatee of Recorde's *Grounde*, became a principal witness at the trial of Lord Protector Somerset. After Somerset's execution he was himself imprisoned in the Tower, from where he encouraged Richard Eden in his experiments in transmutation. In Mary's reign he was elected a member of parliament, although he had no great enthusiasm for the Marian restoration. During 'Queen Mary's dismal days' he dared to entertain the scholar William Ford, 'a great enemy of papism in Oxford' at Welbeck Abbey, his home in Nottinghamshire. Although he suffered several imprisonments for fraud and financial peculations, he survived them all to become a very wealthy man. He died on 23 November 1583 and is buried at St Wilfrid's Church, Screveton. The church contains memorials to him, his three wives and his twenty-three children.

Robert Talbot, Recorde's collaborator in Anglo-Saxon studies, became prebendary at Wells and Norwich cathedrals in turn. He died in 1558, the same year as Recorde.

John Bale, the quarrelsome 'Bilious Bale' who frequented Recorde's house to consult the books in his library, was forced to flee the country on matters of religion when Queen Mary succeeded to the throne. His ship was captured by a Dutch man-of-war, which was driven by bad weather into St Ives, Cornwall. Bale was arrested on suspicion of treason but was soon released. He eventually made his way to the Netherlands and thence to Frankfurt and Basel. During his exile, he devoted himself to writing. After his return, on the accession of Queen Elizabeth I, he received a prebendal stall at Canterbury, where he died in November 1563. He is buried in the cathedral.

William Sharington, Recorde's embezzling predecessor at the Bristol mint, survived imprisonment in the Tower and was eventually pardoned of all his crimes against the Crown. He died a rich man in November 1553. A drawing of him by Holbein now hangs in Windsor Castle.

Recorde's friend and patient Edward Underhill recovered his position as a gentleman-at-arms after his release from Newgate, helped by his defence of Queen Mary during Wyatt's rebellion. He never abandoned his religious zeal, and although a courtier and personal bodyguard to the queen, he allowed groups of Protestant reformers to meet without interference and tried to hinder Mary's growing persecutions by bricking up evangelical tracts behind a wall at his dwelling in Limehouse. No doubt his personal loyalty to Mary and the influence of his many powerful patrons contributed to his unusual standing, as one who retained a position of trust at the court of the Catholic queen and yet used it for fairly undisguised evangelical activities. He spent his final years at Bagginton in Warwickshire and died in 1576.

Richard Eden, the confederate of Recorde's brother in alchemical experiments, escaped unscathed from the charge of heresy which brought him before Stephen Gardiner, bishop of Winchester, probably due to Gardiner's timely death. He became renowned as a translator of many books into English, his most well known being the translation from Spanish of Martin Cortes's *Breve Compendio de la sphaera y de la arte de navigar*. Eden's title was *The Arte of Navigation Conteining a compendious description of the Sphere*, the first English manual on navigation, which formed part of Francis Drake's ship's library on his circumnavigation of the world. Eden died in the early part of 1576.

Richard Chancellor, the seaman who met with Recorde and others at Recorde's house, made a second voyage to Muscovy. He returned with the Russian ambassador, Osip Napeya, on the *Edward Bonaventure*, which reached Scotland but was wrecked in Pitsligo Bay on the Aberdeenshire coast on 10 November 1556. Chancellor saved the ambassador's life by carrying him ashore in the ship's boat through mountainous seas, but was himself drowned when the boat capsized.

Recorde's lawyer friend Robert Lougher became Regius Professor of Civil Law at Oxford University in 1556. In 1572 he was elected member of parliament for Pembroke. He died in 1585 and was buried in Tenby.

Recorde's nemesis the earl of Pembroke was sufficiently trusted by Queen Mary to be named one of the executors of her will. When her successor Elizabeth I entered London he bore the sword before her and clearly had her trust and respect. Pembroke's health deteriorated in the 1560s, but this did not stop his continuing pursuit of wealth and he was as ready as ever to take risks for money. In 1552 he bought shares in the ill-fated expedition led by Sir Hugh Willoughby to search for the North-east Passage to China. Pembroke died aged sixty-three at Hampton Court on 17 March 1570. Elizabeth wrote a letter of condolence to his widow.

Robert Recorde showed great zeal for scholarship and teaching throughout his tragically short life. His efforts to popularise mathematics, the science he believed to be the prerequisite of all knowledge, were sustained, uninterrupted and persistent right up to the final moments of his life. He retained his independent spirit and was always ready to provide those who were 'desirous to learn' with his own particular methods of instruction, even when working as a government official. His constant association with newly emerging groups such as the printers, miners, merchants and seamen enabled him to see that they were truly in need of the knowledge he had worked so long and hard to possess himself. The undoubted success of his books was a matter not of genius but of practical experience. The first scientist of the Tudor age who wished to make his knowledge public property, Recorde was superbly equipped for the task.

His first book on arithmetic, *The Grounde of Artes*, was phenomenally successful and Reyner Wolfe printed six editions during Recorde's lifetime. Incredibly, given all his other preoccupations, Recorde found time to add a treatise on fractions for the 1552 edition. Seven more editions followed after Recorde's death, edited between 1561 and 1579 by his fellow scholar Dr John Dee, who added his augmentations 'to that which my friend hath well begun'. Five more editions between 1582 and 1610 were edited by the Norwich schoolteacher John Mellis. Three editions under the editorship of Robert Norton, who held the position of master gunner of England and was himself an author on artillery and mathematics, appeared between 1615 and 1623. An edition augmented by Robert Hartwell appeared in 1618 and thereafter nine more editions were printed between 1623 and 1654, with supplementary tables added by 'RC'. Thomas Willsford edited four editions between 1658 and 1673. In an address to his readers, Willsford told them that 'you will here receive an old arithmetic from the authority of Recorde, entailed upon the people, ratified and signed by the approbation of time'. A final edition edited by Edward Hatton appeared in 1699, making an amazing total for a textbook of thirty-six editions – some authorities suggest there were more, maybe as many as forty-five – spread over an incredible century and a half. In his preface Hatton said that:

> Though the author of the following treatise was one of the most eminent arithmeticians of his time (as appears by the great variety of compendious and excellent rules therein, and the esteem and credit the book acquired for near 150 years together) yet at length the style and phrase growing obsolete, and some errors, for want of the author's correction in reprinting, having crept in, the booksellers (not willing so choice a piece of arithmetic should be lost for want of a little polishing, the principal parts being extraordinary) were pleased to recommend the performance thereof to me, and I have taken all the care I could to do the author justice.

Subsequent mathematical textbooks were often printed in heavy English blackface type, since this was the font used by Wolfe for the first edition of Recorde's arithmetic. For many years buyers would not countenance any book on mathematics unless it at least looked like his. The *Grounde* gained an astonishing hold on schools in Britain and even found its way across the Atlantic, where it was used in the American colonies. In 1610 the apprentice sailor Richard Norwood was idling aboard a coastal vessel lying at Yarmouth, while the crew were dragging for her lost anchors. With time on his hands, he made an entry in his journal:

Whilst I was in this employment that having by me Recorde's *Arithmetic* given me a little before by my father, I went through it in whole numbers and fractions in some three weeks, but I caught a spice of scurvy by continual sitting, but especially by careless and ill diet living aboard alone.[1]

We may be certain that Doctor Recorde did not intend that anyone would become ill through studying his books. Norwood was later to become a pre-eminent exponent of applied mathematics in navigation, thereby exemplifying Recorde's influence on future generations.

Recorde's next most successful book was *The Urinal of Physick*, which ran to some eleven editions and remained in print for about one hunded and thirty years. The last edition appeared in 1679 under the revised title *The Judgement of Urines*. His book on geometry, *The Pathway to Knowledge*, was published in three editions, dated 1551, 1574 and 1602, a span of about fifty years. His treatise on astronomy, *The Castle of Knowledge*, was published in two editions, the second in 1596, forty years after his death, by which time it was largely obsolete, having been overtaken by Copernican theories. Nevertheless, the prosperous Cheshire landowner William Moreton was sufficiently inspired by the book to have plaster depictions of Destiny and Fortune, copied from the title page of the first edition, placed on the end tympana of the long gallery at his home, Little Moreton Hall. Inscriptions under these depictions, also copied from the book, read 'The wheel of fortune whose rule is ignorance' and 'The sphere of destiny whose rule is knowledge'. The hall is a moated half-timbered manor house built in about 1504–8, situated about four miles south-west of Congleton in Cheshire. Little Moreton Hall is today in the care of the National Trust and is open to visitors. Martin Frobisher carried a copy of the *Castle* in the ship's library on his first voyage in search of the North-west Passage. He paid 10s. in total for the book and a similar work by William Cuningham titled *The Cosmographical Glasse*.

Recorde's least successful book was *The Whetstone of Witte*, which was published only in a single edition. Although containing the 'second part of arithmetic', perhaps algebra was a step too far for his contemporaries.

However Robert Norman, who had mastered the mathematical works of Recorde, evidently did not think so. In his book *The New Attractive* (1581) he wrote that though English mechanics lacked Greek and Latin, 'yet have they in English ... for arithmetic, Recorde's works, both his first and second parts'. Today historians of mathematics find the *Whetstone* the most fascinating of all Recorde's mathematical texts. Not so Nigel in Sir Walter Scott's novel, *The Fortunes of Nigel*. Obliged to rent a room in the house of an old miser, Nigel asks the servant if there are any books in the house which he might read to pass away the time. She returns the answer 'only her master's *Whetstone of Witte*, being the second part of Arithmetic, by Robert Recorde, with the Cossic Practice and Rule of Equation', which promising volume Nigel declined to borrow.

There were numerous citations of Recorde's texts in books by later authors in the fifty years or so following his death. In 1582 Edward Worsop, a London land surveyor, wrote in *A Discoverie of sundrie errours committed by Landmeasurers ignorant of Arithmetike and Geometrie* that 'sundry learned works of the Mathematicals (for such as understand or affect learning) are extant in our vulgar tongue: as ... the works of Doctor Recorde'. In 1588 Recorde was cited by Cyprian Lucar in his 'Lucar Appendix' to his translation of Niccolò Tartaglia's work on ballistics, *Three Bookes ... concerning the Arte of Shooting in Artillerie.*

In 1593 the Cambridge mathematical-instrument maker Thomas Fale described Recorde in his *Horologiographia: The Art of Dialling* as an individual 'who if he had taken the pains could have written of the science of dialling with great commendation'. Dialling is the mathematical science of calculating and laying out the faces of sundials for timekeeping purposes, and it is notable that Recorde did plan to take the 'pains', having described in the *Pathway* one of his proposed works as concerning 'the arte of making of Dials, both for the day and the night, with certain new forms of fixed dials for the noon and other for the stars, which be set in glass windows'.

In the seventeenth century Gerhard Johann Vossiua, professor of rhetoric and chronology at the University of Leyden, wrote of Recorde that his *opere de arte faciandi horologium ab oblivione nomen suum vindicavit* (that his opus on the art of clock making restores his name from oblivion). If Recorde ever wrote such a work, it is no longer extant. Thomas Hylles, in *The Arte of vulgar arithmeticke*, published in 1600, commended Recorde's works and in 1602, the master carpenter Richard More, in *The Carpenters Rule to measure ordinarie Timber*, did not 'forget to name and praise the trio who had opened the path of improvement to the working man, Recorde, Digges and Dee'.

It is, though, Recorde's one-time medical pupil William Bullein, in a citation in his 1562 medical treatise, *The Bulwarke of Defence*, who gives us the closest we are ever likely to get to an obituary for the great educator:

> How well was he seen in tongues, learned in Arts and in Sciences, natural and moral. A father in Physic whose learning gave liberty to the ignorant with his *Whetstone of Witte* and *Castle of Knowledge*, and finally giving place to eliding nature, died himself in bondage of prison. By which death he was delivered and made free, and yet liveth in the happy land, among the Laureate learned, his name was Dr Recorde.

EPILOGUE

The titles of all Recorde's textbooks make it clear they were intended to form an ordered progression of study. It has been said that everyone who studied some aspect of mathematics or astronomy in Elizabethan England would have done so by joining him somewhere along this journey, or by learning from someone who had done so. At the start of the seventeenth century John Mellis described Recorde as 'the only light and chief lodestone unto the vulgar sort of English men in this worthy science [mathematics], that ever writ in our natural tongue'. Recorde's medical treatise also continued to be consulted as a valuable source of self-help diagnosis and treatment. The editor of the last printing in 1679, calling himself ' a certain eminent physician in Queen Elizabeth's days', wrote that 'the antiquity and pains of the author has caused it to be presented again to the press, hoping, with judicious men, it shall receive acceptance.'

Recorde's readers were spread across the whole of English society and he was widely read well into the period which we call the Scientific Revolution. His books ensured that he remained reasonably well known until about the middle of the eighteenth century, when inevitably his texts were superseded by later and more modern writers. Progressively, as the next century advanced, his very existence was all but forgotten.

He was not entirely erased from collective memory however. Augustus de Morgan referred to him in the *Companion to the British Almanac for 1837*, stating that he had 'discovered' in Recorde a 'proto-Copernican' previously unknown. In 1840, James Orchard Halliwell discussed Recorde's work in an article entitled 'The Reception of the

Copernican Theory in England', which appeared in the *Philosophical Magazine*. W. Rouse Ball, in his *History of the Study of Mathematics at Cambridge*, published in 1889, said that Recorde was one of only two mathematicians of any note in the first half of the sixteenth century (the other being the bishop of London, Cuthbert Tunstall). The *Oxford Dictionary of National Biography* recognised Recorde as meriting inclusion with an entry by W. F. Sedgwick in 1896. This biography was unfortunately not entirely accurate, but it was not until 2004 that it was replaced with a new entry by Stephen Johnston.

During the First World War the *Western Mail* pointed out that, regrettably, no man of science was featured among a fine collection of statues of eminent Welshmen unveiled by the prime minister, David Lloyd George, in Cardiff. This provoked a lively correspondence in the columns of *Nature* during November 1916. One letter writer suggested that 'as a statue or two are still to be added' Recorde, although little known to modern Welshmen, might be included so that 'science may yet be represented in the Welsh Valhalla'. This provoked a response from another correspondent, who bemoaned Recorde's lack of fame and likened celebrity 'to the river which submerges merit and floats mediocrity to its destination'. Summarising many of Recorde's achievements he observed that 'it was no wonder that this great Cambrian man of science ... found no time to thimblerig for a knighthood'.

The beginning of the twentieth century saw a great revival of interest in Recorde, especially and unexpectedly by American writers. Many articles about him appeared in American and English journals and he received honourable mention in an increasing number of books concerned with the history of mathematics. The most surprising revival, however, occurred when a reverend gentleman in Harrow went bargain hunting.

William Done Bushell was for fifty years honorary chaplain, housemaster and mathematics teacher at Harrow School. He had connections with Recorde's birthplace, and was lord of the manor of Caldey Island, off the Pembrokeshire coast and close to Tenby. He seems to have been a colourful character who enjoyed dressing in the panoply of a Welsh druid. About 1880 he went to a house clearance sale at the top of Harrow Hill and there, among the furniture and bric-a-brac, he discovered a small and very dirty portrait painted on a wooden panel. What must have truly astonished him was that the panel bore the painted inscription 'Robt Recorde M. D. 1556'. It appears that Bushell accepted the painting as a genuine portrait of Recorde solely on the strength of the inscription and without any further enquiry.

No one, at the time or subsequently, seems to have reflected on the staggering coincidence of the painting turning up, of all the places in Britain where it might have surfaced, in Harrow, a town where there happened to reside an eccentric mathematician with connections to Tenby, who was one of only a handful of people in the country at that time to whom the name of Recorde would have meant anything. It is surely pushing credulity too far to suppose that serendipity brought Bushell and the painting together. The most likely explanation is that he was the victim of a hoax, a practical joke perpetrated perhaps by a junior master or even by the boys of his house. It would have required little expertise to paint a convincing inscription on the portrait and plant it where the gullible housemaster would find it.

Bushell died in 1917, but the portrait lived on to confuse and confound later generations. Dr R. T. Gunther, of Magdalen College, Oxford, borrowed it from Bushell's widow for display at the Ashmolean Museum, in an exhibition of early scientific instruments that ran from 1920 to 1925. This prestigious event, during which the authenticity of the painting was again not queried, lent the portrait credibility by association. Gunther caused a picture postcard of the painting to be made for sale during the exhibition, identified as No 41 in the Old Ashmolean Series. He also reproduced the portrait in his *Early Science at Oxford*, published in 1923, thus lending the picture further respectability.

Bushell's eldest son inherited the painting and in 1940 he gifted it to the Faculty of Mathematics at Cambridge University, as a fitting memorial to the great scholar who had once studied there. However, he had begun to have doubts about the picture and before handing it over, he left it for three weeks at the National Portrait Gallery in London for an opinion as to its authenticity.[1] With no attribution or provenance, there was little to go on and Sir Henry Hake, the director of the NPG, and his team of experts were noncommittal in their judgement. They pointed out that the picture had been cleaned, over-cleaned and touched up with paint. The sitter's gown had been cleaned almost out of existence and only the white collar had any character. The board on which it was painted, they said, may well have been cut from a larger painting. Their most damaging conclusion, however, was that the inscription was painted over the dirt on the surface, late in the nineteenth century and probably in the 1880s. This coincides exactly with the date of the suspected hoax.

Despite it being almost certain that the painting was bogus, the Faculty of Mathematics accepted it and, with the Second World War now raging, it was put away in a cupboard. In 1946 the painting was recovered from its wartime hiding place and displayed in the Philosophical Library. Its existence became known to All Souls College, Oxford, which expressed deep interest and admitted that Dr Recorde had been totally forgotten by his old college. The warden arranged for a photograph of the portrait to be put up in the Writing Room, although it is not certain if this was ever done.

In 1964 the Department of Pure Mathematics and Mathematical Statistics was established at Cambridge and the existence of the painting was recalled, it having apparently been put away again. Sir William Hodge, the first head of the department, had it brought out and cleaned (again!) with a view to hanging it in his office. For some unexplained reason it remained in its wrappings in a cupboard after this cleaning, until in 1969 Professor J. Cassels took over as head of department. On examining the portrait he was greatly surprised to find that the 'Robt Recorde' inscription had disappeared, presumably removed with the surface dirt during its last cleaning. However, and most significantly, a faint *Aetat. suae 63 A° 1631* could just be made out where the inscription had previously been. *Aetat.* is an abbreviation of the Latin *Aetatis* meaning 'age' and the whole phrase can be interpreted as '63 years of age'. *A° 1631* is shorthand for *Anno Domini 1631*, the year the picture was painted. Mr David Piper of the Fitzwilliam Museum made an examination and his verdict damns the authenticity of the picture for all time:

> To the naked eye and under a glass the revealed inscription appears to be coeval with the rest of the paint. I would think the picture is

Flemish and I see no reason to query the date of 1631 which is in perfect accordance with the costume. The latter suggests a Flemish rather than an English sitter, and I can see no particular reason for this painting having been done in England. This, alas, rules out I fear the identification as Robert Recorde.

FIGURE 22 The false portrait of Robert Recorde.

The pernicious portrait on the right, 'discovered' by William Done Bushell and once supposed to be Recorde's likeness, is now long discredited. The medallion memorial in St Mary's Church, Tenby, sculpted by Owen Thomas and based on the false portrait, is illustrated on the left with its misleading inscription.

In the first decade of the twentieth century another strand was added to the tale of this pernicious portrait. It was decided, by whom no one now seems to know, that Recorde's old place of worship, St Mary's in Tenby, should contain a memorial to celebrate the four-hundredth anniversary of his birth. Accordingly a local sculptor, Owen Thomas, visited the Bushell home in Harrow for the purpose of making sketches, the painting still being in the possession of the family at that time. In 1910 a portrait medallion was sculpted, with a likeness based on the Harrow painting. The memorial was placed in St Thomas's Chapel, with an inscription:

> In memory of Robert Recorde, the eminent mathematician, who was born at Tenby, circa 1510. To his genius we owe the earliest important English treatises on algebra, arithmetic, astronomy and geometry; he also invented the sign of equality = now universally adopted by the civilized world. Robert Recorde was court physician to King Edward VI and Queen Mary. He died in London, 1558.

The inscription is misleading. Apart from his incorrect birth date, there is no evidence whatsoever that Recorde was ever consulted on medical matters by any Tudor monarch. To remedy the situation, Tenby Museum placed an information board giving the correct facts below the memorial. It is perhaps time, after more than one hundred years during which Recorde has suffered the indignity of having another man's face masquerading as his own, to have this memorial consigned to a vault as a historical curiosity. It could more fittingly be replaced by a dignified slab of Welsh slate, engraved with accurate biographical details, something this latter-day icon of Welsh nationality surely deserves.

Another twist to the portrait saga was added in 1958 at a meeting of the south-west Wales branch of the Mathematical Association, held to commemorate the four-hundredth anniversary of Recorde's death. A painting was displayed, similar in appearance to the Harrow portrait and seemingly by an amateur painter, copied from the portrait medallion in Tenby church. Delegates were dismayed to be told categorically by Mr Prag of Westminster School, who was aware of the discredited portrait at Cambridge, that this was not a portrait of the Welsh polymath. The artist may have been a student of Trinity College Carmarthen, where this 'second' portrait is now archived. The Trinity portrait was brought out again and shown at a week-long conference at Gregynog Hall in Powys, held to mark the four-hundred and fiftieth anniversary of Recorde's death.

Today the internet is awash with images of the bogus picture, as a Google search will quickly reveal. After this length of time it seems that Recorde can never be dissociated from the picture, which probably depicts a foreign national who flourished seventy-three years after his death, and whose age exceeded that attained by him by some fifteen years. Sadly, it seems that it has to be accepted that people will for evermore continue to regard the face of an unknown man as his likeness.

A much more fitting memorial to Recorde than the one in Tenby church is located in the Department of Computer Science at Swansea University. The department has named its seminar and conference room 'The Robert Recorde Room' and a large plaque, commissioned in 2001, is situated by the entrance. It was designed by the artist John Howes and carved by the calligrapher Ieuan Rees on Welsh slate. The quotation reproduced on it comes from Recorde's *The Castle of Knowledge* and reads:

> in all men's works, you be not abused by their authority, but evermore attend to their reasons, and examine them well, ever regarding more what is said, and how it is proved, then who says it, for authority often times deceives many men.

Tenby Museum currently maintains an exhibition celebrating Recorde's life and works, and some early editions of his books are permanently displayed. To commemorate the five-hundredth anniversary of his birth a banner was erected across Tenby High Street and memorial lectures were hosted by the museum. An art exhibition, '2010–1510 = Robert Recorde', featured fifty artists, who displayed imaginative and ingenious interpretations of his life and notable achievements. All these events were well attended and serve to indicate that Recorde is now a respected historical figure, of whom the general public is no longer in complete ignorance.

A final postscript to Recorde's story concerns the ubiquitous symbol for equality that he devised more than five hundred years ago. With the advent of computer algebra systems in the twenty-first century, mathematicians became aware that a single symbol like Recorde's = was no longer adequate for every case. Nowadays it is necessary to distinguish between the sort of equality that is always true, as in $2 + 2 = 4$, and the sort that is true only for a limited time by virtue of a definition, which requires a new symbol. During the writing of a computer programme, we might define an algebraic variable, say n, as being equal to some particular number. A common way of doing this in current computer systems is to modify Recorde's symbol by adding a colon and writing, for example, $n := 4$, and then later in the program $n := 6$, and so on for any other value of n we care to define. In summary, Recorde's symbol means equal forever, while the computer symbol means equal only until the equality is redefined. The $:=$ symbol is not yet universally accepted and only time will tell whether it becomes as longlasting as Recorde's two parallel lines, or is replaced by something else.

No doubt Recorde would have regarded all this as a 'strange art', but he laid down the principles of teaching difficult subjects, like modern computer algebra, long before Elizabeth I came to the throne. In his geometry textbook, *The Pathway to Knowledge*, he tells us that:

> it is not easy for a man that shall travail in a strange art, to understand at the beginning both the thing that is taught and also the just reason why it is so. And by experience of teaching I have tried it to be true, for when I have taught the proposition, as its import in meaning, and annexed the demonstration withal, I did perceive that it was a great trouble and a painful vexation of mind to the learner, to comprehend both those things at once. And therefore did I prove first to make them to understand the sense of the propositions, and then afterward did they conceive the demonstrations much sooner, when they had the sentence of the propositions first inserted in their minds. This thing caused me ... to omit the demonstrations, and to use only a plain form of declaration, which might best serve for the first introduction ... for so shall men best understand things, first to learn that such things are to be wrought, and secondarily what they are, and what they do import, and then thirdly what is the cause thereof.[2]

We cannot forget that 'God of his mercy' ended Robert Recorde's troubles soon enough in death, but his gentle voice lives on through the centuries in his writings, evidence enough of the passing not only of a kindly spirit but, surely, a glorious one.

NOTES AND REFERENCES

Prologue

1. Robert Recorde, *Tenby's Famous Renaissance Scholar*. Undated pamphlet issued by Tenby Museum and Art Gallery. The source of the quoted pen portrait is now lost.
2. Thavit Sukhabanji, 'Mathematical Messiah, Robert Recorde and the Popularization of Mathematics in the Sixteenth Century' (unpublished MA thesis, North Texas State University, Denton, Texas, 1980), 10.
3. Anthony à Wood, *Athenæ Oxonienses: An Exact History of all the Writers and Bishops who have had their Education in the University of Oxford*, with additions by Philip Bliss, (London: F. C. and J. Rivington and others, 1813), vol. 1, cols. 255–6.
4. Samuel Rush Meyrick, *Heraldic Visitations of Wales and Part of the Marches, by Lewys Dwnn, Deputy Herald at Arms* (Llandovery: William Rees, 1846), vol. 1, pp. 68–9. Mr J. M. A. Petrie, Rouge Croix Pursuivant, College of Arms, carried out an archival search on behalf of the author, but was unable to discover anything more than is contained in Meyrick's publication. This is not considered an official record by the College of Arms and accordingly it cannot be confirmed that the Recorde family had a right to bear arms.
5. Chris Skidmore, *Bosworth* (London: Weidenfeld and Nicolson, 2013), p. 29.

Chapter 1

1. Anthony à Wood, *Athenæ Oxonienses: An Exact History of all the Writers and Bishops who have had their Education in the University of Oxford*, with additions by Philip Bliss (London: F. C. and J. Rivington and others, 1813), vol. 1, cols. 255–6.
2. A three-storey Tudor merchant's house has been preserved unaltered in Tenby. Now in the care of the National Trust, the house and shop have been furnished to show how it may have looked around the time of Robert Recorde's birth; see <http://www.nationaltrust.org.uk/tudor-merchants-house>.
3. Robert Recorde, *The Grounde of Artes* (London: R. Wolfe, 1543) pp. 88–96.
4. Recorde, *The Grounde of Artes*, pp. 134–7.
5. Services in the chapel had to be cancelled when rough weather caused waves to break over the quay. The building, in later days often referred to as the Fisherman's Chapel, was demolished in 1840 in order to widen the harbour entrance. It was replaced by the

present day seaman's chapel of St Julian, erected in a more sheltered location on the landward side of the harbour.
6. Herbert F. Hore, 'Mayors and Bailiffs of Tenby', *Archaelogia Cambrensis*, 8 (1853), 117.

Chapter 2

1. Anthony à Wood, *Athenæ Oxonienses: An Exact History of all the Writers and Bishops who have had their Education in the University of Oxford*, with additions by Philip Bliss (London: F. C. and J. Rivington and others, 1813), vol. 1, cols. 255–6.
2. Wood, *Athenæ Oxonienses*, vol. 2, col. 84.
3. Wood, *Athenæ Oxonienses*, vol. 1, cols. 255–6.
4. John Venn (ed.), *Grace Book, Containing the Records of the University of Cambridge, 1542–1589* (Cambridge: Cambridge University Press, 1910), p. 2.
5. The conditions for the award of an Oxford Licentiate in Medicine can be found in Charles Edward Mallet, *A History of the University of Oxford* (London: 1924), vol. 2, pp. 84–5.
6. Wood, *Athenæ Oxonienses:* vol. 1, cols. 255–6.
7. Wood, *Athenæ Oxonienses*: vol. 1, cols. 261–2.
8. Susan Wabuda, 'Thomas Garrard, Clergyman and Protestant Reformer', *Oxford Dictionary of National Biography*, (Oxford: Oxford University Press, 2004); online edn (January 2008), see <http://www.oxforddnb.com/view/article/10560>.
9. R. T. Gunther, *Early Science in Oxford* (Oxford: Oxford University Press, 1937), vol. 11, pp. 60, 156.

Chapter 3

1. Charlotte Fell-Smith, *John Dee* (London: Constable, 1909), pp. 241–2.
2. Robert Recorde, *The Urinal of Physick* (London: Reynolde Wolfe, 1547), sig. Av.
3. Alan Bryson, 'Richard Whalley, Administrator', *Oxford Dictionary of National Biography* (Oxford: Oxford University Press, 2004); online edn (January 2008), see <http://www.oxforddnb.com/view/article/29161>
4. Travis D. Williams, 'The Earliest English Printed Arithmetic Books', *The Library: Transactions of the Bibliographical Society*, 13/2 (June 2012), pp. 164–84.
5. A. W. Richeson, 'The First Arithmetic Printed in English', *Isis*, 37/1–2 (May 1947), 47–56.

Chapter 4

1. 'Blynne' is a word now archaic but obviously familiar in Recorde's time. Its roots probably lie in Middle English *blenchen*, meaning 'flinch'.

Chapter 5

1. H. S. Bennett, *English Books and Readers 1475–1557* (Cambridge: Cambridge University Press, 1969), p. 35.

2. Andrew Pettegree, 'Reyner Wolfe, Printer and Bookseller', *Oxford Dictionary of National Biography*, (Oxford: Oxford University Press, 2004), online edn (September 2011), see <http://www.oxforddnb.com/view/article/29835>.
3. E. Gordon Duff, *A Century of the English Book Trade* (New York: Cambridge University Press, 1905), pp. 171–2.
4. Frederick Wilson and Douglas Grey, *A Practical Treatise upon Modern Printing Machinery and Letterpress Printing* (London: Cassell, 1888), pp. 3–8.

Chapter 6

1. John Venn (ed.), *Grace Book, Containing the Records of the University of Cambridge, 1542–1589* (Cambridge: Cambridge University Press, 1910), p. 530.
2. Venn, *Grace Book,* p. 27.
3. David Eugene Smith and Frances Marguerite Clarke, 'New Light on Robert Recorde', *Isis*, 8/1 (February 1926), 67.
4. National Archives, Kew, 'Will of Robert Recorde, Doctor of Physic', PROB 11/40/317.

Chapter 7

1. H. S.Bennett, *English Books and Readers 1475–1557* (Cambridge: Cambridge University Press, 1969), p. 36.
2. National Archives, Kew, 'Will of Joan Wolfe, Widow of City of London', PROB 11/56/386. The will comprises more than six pages and gives much valuable information about the Wolfe family.
3. John Stowe, *A Survey of London*, (Oxford: Clarendon Press, 1908), reprinted from the text of 1603, vol. 1, pp. 293, 330.
4. See Timothy Graham, 'Anglo-Saxon studies: sixteenth to eighteenth centuries', in Phillip Pulsiano and Elaine Treharne (eds), *A Companion to Anglo-Saxon Literature* (Oxford: Blackwell, 2001), pp. 415–18.
5. Annotated by John Bale, writing in Latin, as 'Ex Museo Robertus Recorde'.
6. I am indebted to Sue Baldwin, Honorary Librarian of Tenby Museum and Art Gallery, for providing me with a list of the contents of Recorde's library, as reported by Bale and Reginald Poole. Unfortunately the list is long and cannot be reproduced here.
7. Robert Recorde, *The Pathway to Knowledge* (London: Reynold Wolfe, 1551), sig. D–Di.
8. Dale Hoak, 'Edward VI, King of England and Ireland', *Oxford Dictionary of National Biography* (Oxford: Oxford University Press, 2004); online edn (May 2014), see <http://www.oxforddnb.com/view/article/8522>.
9. Recorde, *The Pathway to Knowledge*, sig. av.

Chapter 8

1. John Strype, *Ecclesiastical Memorials relating chiefly to Religion and the Reformation of it* (Oxford: Clarendon Press, 1822), vol. 12, pt. 1, pp.176–81.

2. 'My very friend Mr Recorde, doctor of physick, singularly seen in all the seven sciences, and a great divine', in John Gough Nichols (ed.), *Narratives of the Days of the Reformation* (London: Camden Society, 1859), pp. 150–1.
3. Robert Recorde, *The Castle of Knowledge*, (London: Reginalde Wolfe, 1556), preface.
4. John Gough Nichols (ed.), *Literary Remains of King Edward the Sixth* (London: J. B. Nichols and Sons, 1857), vol. 1, p. cviii.

Chapter 9

1. Jack Williams, *Robert Recorde, Tudor Polymath, Expositor and Practitioner of Computation* (London: Springer, 2011), pp. 20–3.
2. Jack Williams, 'The Lives and Works of Robert Recorde' (unpublished essay, March 2009), p. 3.
3. Robert Recorde, *The Grounde of Artes* (London: R. Wolfe, 1543) pp. 88–91.
4. Sharington was paid an annual salary of £133 6s. 8d during his tenure at the mint. It is reasonable to suppose that Recorde received a similar sum when he assumed the comptrollership.
5. C. E. Boucher, 'St Peter's Church, Bristol', *Transactions of the Bristol and Gloucestershire Archaeological Society*, (Bristol, 1909) vol. 32, p. 290.
6. John Maclean, 'Notes on the Accounts of the Procurators, or Churchwardens, of the Parish of St Ewen's, Bristol – 1', *Transactions of the Bristol and Gloucestershire Archaeological Society* (Bristol, 1890–1), vol. 15, p. 140, n. 2.
7. William Barrett, *The History and Antiquities of the City of Bristol* (Bristol: William Pine, 1787), p. 440.

Chapter 10

1. Jack Williams, *Robert Recorde, Tudor Polymath, Expositor and Practitioner of Computation* (London: Springer, 2011), p. 26.
2. Stephen Reed Cattley (ed.), *The Acts and Monuments of John Foxe* (London: R. B. Seeley and W. Burnside, 1838), vol. 6, pp. 1556.
3. Robert Recorde, *The Castle of Knowledge* (London: Reginalde Wolfe, 1556), p. 133.
4. Harleian MS 167, ff. 106r–108r (London, British Library). This manuscript appears to be a memorandum, headed 'Concerning a passage to be made from our North Sea into the South Sea' and is indirectly dated 12 December 1586. It contains a transcript of two letters by Philip Jones (identified as P. E. Jon), which mention Cabot, Chancellor, Pinteado and Recorde as meeting at his house. According to some sources this document is supposed to contain marginal 'opinions' by these four, but examination of the document on the author's behalf failed to reveal any such notes.
5. Recorde, *The Castle of Knowledge*, pp. 70, 82.
6. Robert Recorde, *The Whetstone of Witte* (London: Jhon. Kyngstone, 1557), sig. aiiiv–aiiir.

Chapter 11

1. According to Johnson and Larkey, Recorde 'handles the conversation with the touch of an artist, and one is conscious of two different people talking.' See Francis R. Johnson and Stanford V. Larkey, 'Robert Recorde's Mathematical Teaching and the Anti-Aristotelian Movement', *Huntingdon Library Bulletin,* 7 (April 1935), 59–87.
2. Robert Recorde, *The Castle of Knowledge* (London: Reginalde Wolfe, 1556), p. 165.

Chapter 12

1. Anthony à Wood, *Athenæ Oxonienses: An Exact History of all the Writers and Bishops who have had their Education in the University of Oxford*, with additions by Philip Bliss (London: F. C. and J. Rivington and others, 1813), Vol. 1, col. 256, n. 7. The verse in question reads:

   > The godly use of prudent witted men
   > Cannot abstain their ancient exercise
   > Recorde of late how busily with his pen
   > The translator of this said treatise
   > Hath him endeavoured in most goodly wise
   > Books to translate in volumes large and faire
   > From French in prose of ghostly exemplaire.

2. Robert Fabyan, *The New Chronicles of England and France* (London: Henry Ellis, 1811), p. 19, n.
3. Fabyan, *The New Chronicles of England and France*, p. 30, n.
4. Howell A. Lloyd, *The Gentry of South-West Wales* (Cardiff: University of Wales Press, 1968), p. 73. In a later reference (p. 153) Lloyd refers to Robert Recorde being involved in a brawl within the liberties of Tenby over wrongful intrusion on his property and being arrested and clapped in irons in the town gaol. This is unlikely to have been the distinguished Dr Recorde, but was probably his young nephew Robert.
5. J. Payne Collier, *A Bibliographical and Critical Account of the Rarest Books in the English Language*, vol. 3, pp. 287–8.

Chapter 13

1. Philip Herbert Hore, *History of the Town and County of Wexford* (London: Elliot Stock, 1901). Much of what is known about the 'Clonmines affair' is summarised in Chapter 3, 'The chronicles of Clonmines', pp. 233–46.
2. Edward Kaplan, 'Robert Recorde (*c.*1510–1558): Studies in the life and works of a Tudor scientist' (unpublished PhD thesis, New York University, 1960), 20.
3. David Eugene Smith, and Frances Marguerite Clarke, 'New Light on Robert Recorde', *Isis*, 8/ 1 (February 1926), 59–63.
4. Smith and Clarke, 'New Light on Robert Recorde', *Isis*, 8/1, 64–5.

Chapter 14

1. Edward Kaplan, 'Robert Recorde (*c*.1510–1558): Studies in the life and works of a Tudor scientist' (unpublished PhD thesis, New York University, 1960), 25.
2. See Chapter 6, 'Mr Underhill's suffering for religion', in John Strype, *Ecclesiastical Memorials relating chiefly to Religion and the Reformation of it* (Oxford: Clarendon Press, 1822), vol. 3, pt. 1, pp. 92–103.
3. Robert Recorde, *The Castle of Knowledge* (London: Reginalde Wolfe, 1556), sig. aii.
4. Jack Williams, email to the author, 10 June 2015. I am indebted to him for pointing out the possibility that Recorde was here signposting the beginnings of algebra.

Chapter 15

1. James Halliwell, *Science in England* (London: R. and J. E. Taylor, 1841), p. x.
2. Robert Recorde, *The Whetstone of Witte*, (London, Jhon Kyngstone, 1557), sig. Sii.
3. Jack Williams, *Robert Recorde, Tudor Polymath, Expositor and Practitioner of Computation*, (London: Springer, 2011), p. 16.

Chapter 16

1. Robert Recorde, *The Castle of Knowledge* (London: Reginalde Wolfe, 1556), p. 284. He prevaricated, however, by adding 'but that fruit will I reserve for another place.
2. David Eugene Smith and Frances Marguerite Clarke, 'New Light on Robert Recorde', *Isis*, 8/1 (February 1926), 67.
3. National Archives, Kew, 'Will of Robert Recorde, Doctor of Physic', PROB 11/40/317.

Chapter 17

1. Wesley Craven and Walter Hayward (eds), *The Journal of Richard Norwood*, (New York: Ann Arbor, 1945), p. 16.

Epilogue

1. I am grateful to Sue Baldwin, honorary librarian, Tenby Museum and Art Gallery, for allowing me to see correspondence, dated 1940 to 1958, between the museum, W. D. Bushell, Lionel Curtis of All Souls and others, concerning the history and verification of the portrait.
2. Robert Recorde, *The Pathway to Knowledge* (London: Reynold Wolfe, 1551), second book, preface, sig. a.iiv.

SELECT BIBLIOGRAPHY

This bibliography is given as a starting point for further research. It includes all the extant works of Robert Recorde and all works cited in the text, except a few that do not have a general relevance to the themes of the book. It also includes works and articles not cited but relevant to the life and times of Recorde.

Agricola, Georgius, *De Re Metallica* (New York: Dover Publications, 1950), translated from the first Latin edition of 1556 by H. and L. Hoover).

Anstey, Henry, *Academical Life and Studies at Oxford*, pt. 1 (London: Longmans, Green, Reader and Dyer, 1868).

Ashmole, Elias, *Theatrum Chemicum Britannicum* (London: Nath. Brooke, 1652).

Baron, Margaret E., 'A Note on Robert Recorde and the Dienes Blocks', *Mathematical Gazette*, 50/374 (December 1966), 363–9.

Barr, William, 'A World View of Robert Recorde: A Brief Study of Tudor Cosmology', *Albion: A Quarterly Journal Concerned with British Studies*, 1/1 (1969), 1–9.

Barrett, William, *History and Antiquities of the City of Bristol* (Bristol: William Pine, 1789).

Bennett, H. S., *English Books and Readers 1475–1557* (Cambridge: Cambridge University Press, 1969).

Berkenhout, John, *Biographical History of Literature*, vol. 1 (London: J. Dodsley, 1777).

Besant, Walter, *The History of London*, (London: Longmans, Green and Co., 1894).

———, *London in the Time of the Tudors* (London,: Adam and Charles Black, 1904).

Bullein, William, *Bullein's Bulwarke of Defence Against all Sickness, Sornes and Wounds* (London: John Kyngston, 1562).

Chalmers, Alexander, *The General Biographical Dictionary of the Most Eminent Persons in Every Nation*, vol. 26 (London: J. Nichols and Son, 1816)

Collier, J. Payne, *A Bibliographical and Critical Account of the Rarest Books in the English Language*, vol. 3 (New York: Francis and Scribner, 1866).

Cook, Harold J., 'Good Advice and Little Medicine: The Professional Authority of Early Modern English Physicians', *Journal of British Studies*, 33/1 (January 1994), 1–31.

De Morgan, A., *Arithmetical Books from the Invention of Printing to the Present Time* (London: Taylor and Walton, 1847).

Duff, E. Gordon, *A Century of the English Book Trade* (New York: Cambridge University Press, 1905).

Easton, Joy B., 'On the date of Robert Recorde's birth', *Isis*, 57/1 (Spring, 1966), 121.

———, 'The Early Editions of Robert Recorde's Ground of Artes', *Isis*, 58/4 (Winter 1967), 515–32.

Elton, G. R., *England under the Tudors* (Abingdon: Routledge, 1991).

Fabyan, Robert, *The New Chronicles of England and France*, two parts, reprinted from Richard Pynson's edition of 1516 (London: Henry Ellis, 1811).

Fell-Smith, Charlotte, *John Dee* (London: Constable, 1909).

Foster, Joseph (ed.), *Alumni Oxonienses: The Members of the University of Oxford, 1500–1714* (London, 1891).

Goodison, J. W., *Catalogue of Cambridge Portraits* (Cambridge: Cambridge University Press, 1955).

Griffiths, George, *Chronicles of the County of Wexford* (Enniscorthy: Watchman's Office, 1877).

Gunther, R. T., *Early Medical and Biological Science* (Oxford University Press and London: Humphrey Milford, 1926).

———, *Early Science in Oxford* (Oxford: Oxford University Press, 1937).

Halliwell, James, *The Connection of Wales with the Early Science of England* (London: Richard and John E. Taylor, 1840).

———, *Science in England* (London: R. and J. E. Taylor, 1841).

Hamilton, Hans Claude (ed.), *Calendar of the State Papers of Ireland 1509–1573*, (London: Longman, Green, Longman and Roberts, 1860).

Hore, Philip Herbert, *History of the Town and County of Wexford* (London: Elliot Stock, 1901).

Hutton, Charles, *Tracts on Mathematical Subjects*, vol. 2 (London: C. and J. Rivington, 1812).

Johnson, Francis R. and Larkey, Stanford V., 'Robert Recorde's Mathematical Teaching and the Anti-Aristotelian Movement', *Huntingdon Library Bulletin*, 7 (April 1935).

Johnston, Stephen, 'Recorde, Robert (*c.*1512–1558)' *Oxford Dictionary of National Biography* (Oxford University Press, 2004).

Ives, Eric, *The Reformation Experience* (Oxford: Lion Hudson, 2012).
Kaplan, Edward, 'Robert Recorde (c.1510–1558): Studies in the life and works of a Tudor scientist' (unpublished PhD thesis, New York University, 1960).
Lloyd, Howell A., *The Gentry of South-West Wales, 1540–1640* (Cardiff: University of Wales Press, 1968).
———, 'Famous in the Field of Number and Measure: Robert Recorde, Renaissance Mathematician', *Welsh History Review*, 2 (2000), 254–82.
Mayers, Kit, *North-East Passage to Muscovy* (Stroud: Sutton Publishing, 2005).
McConica, J. K., 'The Social Relations of Tudor Oxford' *Transactions of the Royal Historical Society*, fifth series, 27 (1977), 115–34.
McKisack, May, *Medieval History in the Tudor Age* (Oxford: Clarendon Press, 1971).
Meyrick, Samuel Rush (ed.), *Heraldic Visitations of Wales and Part of the Marches*, by Lewys Dwnn, Deputy Herald at Arms (Llandovery, William Rees, 1846).
Morley, Frank V., 'Finis Coronat Opus', *The Scientific Monthly*, 10/3 (March 1920), 306–8.
Nichols, John Gough (ed.), *Literary Remains of King Edward the Sixth* (London: J. B. Nichols and Sons, 1857).
———, *Narratives of the Days of the Reformation*, (London, The Camden Society, 1859).
O'Malley, C. D., 'Tudor Medicine and Biology', *Huntington Library Quarterly*, 32/1 (November 1968), 1–27.
Owen, George, *The Description of Pembrokeshire* (1603), reprinted in the Welsh Classics series (Llandysul: Gomer Press, 1994).
Page, R. I., 'Anglo-Saxon Texts in Early Modern Transcripts', *Transactions of the Cambridge Bibliographical Society*, 6// 2 (1973), 69–85.
Patterson, Louise Diehl, 'Recorde's Cosmography, 1556', *Isis*, 42/3 (October 1951), 208–18.
Pugh, Ralph B., *Imprisonment in Medieval England* (Cambridge: Cambridge University Press, 1968).
Pulsiano, Phillip, and Treharne, Elaine (eds), *A Companion to Anglo-Saxon Literature* (Oxford: Blackwell, 2001).
Rait, Robert S., *Life in the Medieval University* (London: Cambridge University Press, 1912).
Recorde, Robert, *The Grounde of Artes*, (London: R. Wolfe, 1543), facsimile of the 1543 first edn, (Derby: TGR Renascent Books, HB 2009, PB 2012).
———, *The Urinal of Physick*, (London: Reynolde Wolfe, 1547), facsimile of the 1547 first edn, (Derby: TGR Renascent Books, HB 2011, PB 2012).
———, *The Pathway to Knowledge*, (London: Reynold Wolfe, 1551), facsimile of the 1551 first edn, (Derby: TGR Renascent Books, HB 2009, PB 2013).

———, *The Castle of Knowledge*, (London: Reginalde Wolfe, 1556), facsimile of the 1556 first edn, (Derby: TGR Renascent Books, HB 2009, PB 2012).

———, *The Whetstone of Witte*, (London, Jhon Kyngstone, 1557), facsimile of the 1557 first edn, (Derby: TGR Renascent Books, HB 2010, PB 2013).

Richardson, W. C., 'Some Financial Expedients of Henry VIII', *Economic History Review*, New Series, 7/1 (1954), 33–48.

Ridley, Jasper, *The Tudor Age* (London: Constable, 1998).

Roberts, Gareth, and Smith, Fenny (eds), *Robert Recorde, The Life and Times of a Tudor Mathematician*, (Cardiff: University of Wales Press, 2012).

Rouse Ball, W. W., *A History of the Study of Mathematics at Cambridge* (Cambridge: Cambridge University Press, 1899).

Sedgwick W. F., 'Robert Recorde, Mathematician', *Dictionary of National Biography*, vol. 47 (Oxford: Oxford University Press, 1896).

Sleight, E. R., 'Early English Arithmetics', *National Mathematics Magazine*, 16/4 (January 1942), 198–215 and 16/5 (February 1942), 243–51.

Smith, David Eugene, 'New Information Respecting Robert Recorde', *American Mathematical Monthly*, 28/8–9 (August–September, 1921) 296–300.

———, and Clarke, Frances Marguerite, 'New Light on Robert Recorde', *Isis*, 8/1 (February 1926), 50–70.

Smith, Goldwin, 'The Practice of Medicine in Tudor England', *Scientific Monthly*, 50/1 (January 1940), 65–72.

Strype, John, *Ecclesiastical Memorials relating chiefly to Religion and the Reformation of it* (Oxford: Clarendon Press, 1822), vols 1, 2 and 3.

Sukhabanji, Thavit, 'Mathematical Messiah, Robert Recorde and the Popularization of Mathematics in the Sixteenth Century' (unpublished MA thesis, North Texas State University, Denton, Texas, 1980).

Venn, John (ed.), *Grace Book, Containing the Records of the University of Cambridge, 1542–1589* (Cambridge: University of Cambridge Press, 1910).

Whittle, C. H., 'Medicine At Cambridge', *British Medical Journal*, 1/4556 (May 1948), 111–113.

Williams, Jack, *Robert Recorde, Tudor Polymath, Expositor and Practitioner of Computation* (London: Springer, 2011).

Wilson, Frederick, and Grey, Douglas, *A Practical Treatise upon Modern Printing Machinery and Letterpress Printing* (London: Cassell, 1888).

Wood, Anthony à, *Athenæ Oxonienses: An Exact History of all the Writers and Bishops who have had their Education in the University of Oxford*, with additions by Philip Bliss (London: F. C. and J. Rivington and others, vol. 1, 1813, vol. 2, 1815).

Zetterberg, J. Peter, 'The Mistaking of "the Mathematicks" for Magic in Tudor and Stuart England', *Sixteenth Century Journal*, 11/1 (Spring, 1980), 83–97.

INDEX

Abell, Thomas, 91
Allen, Richard (the false prophet), 87, 88, 89
An introduction for to learn to reckon, 35
Art of numbering by the hand, 13

Bale, John, 80, 191
Barnes, Robert, 90
Bayfield, Richard, 56
Berthelet, Thomas, 57
Bilney, Thomas, 25
Bloomfield's Blossoms, 133
Boleyn, Anne (Queen of England), 57
Book burning, 60
Book censorship, 60
Boreman, Richard, 36
Bullein, William, 64, 197
Bullinger, Heinrich, 56
Bushell, Rev William Done, 200

Cabot, Sebastian, 109
Calicut, 112
Cambridge University
    distinguished community of scholars, 177
    medical studies, 29–32
    St John's College, 27, 34
Cape of Good Hope, 121
Chamberlain, Sir Thomas, 101, 102
Chancellor, Richard, 109, 114, 120, 192
Cheke, Sir John, 85, 110
Clonmines, Ireland, 144, 148

Collier, J. Payne, 139
Copernicus, Nicolaus, 125
Cranmer, Thomas
    appointed Archbishop of Canterbury, 55
    brings Reyner Wolfe to England, 55
    tried and condemned for heresy, 158
Croft, Sir James, 144, 153
Cum privilegio ad imprimendum solum, 75

De Auricula Confessione, 92
De negotio Eucharistie, 92
Debasement of coinage, 97
Dee, John, 27, 190, 193
Dudley, Sir John (duke of Northumberland), 108, 144, 154, 155
Dwnn, Lewys, 4
Dynbech-y-pyscoed, 7

Eden, Richard, 132, 173, 192
Evenet, Giles, 102

Fabyan, Robert, 129
Featherstone, Richard, 91
Fetcher or bringer of scholars, 17
First English arithmetics, 35
Frobisher, Martin, 111, 195
Froschover, Christopher, 56

Galena and silver deposits, 143
Gardiner, Stephen (Bishop of Winchester)
  brought to trial at Lambeth, 109
  imprisonment in the Tower of London, 96
  preaches public sermon, 95
  released from prison by Queen Mary, 155
Garrard, Thomas
  book smuggler, 23
  escapes from Oxford, 23
  his death by burning, 90
Geoffrey of Monmouth, 129
Gough, John, 57
Gresham, Sir John, 87
Grey, Lady Jane, 155, 158
Grynaeus, Simon, 55, 124
Gundelfinger, Joachim (German mining captain)
  arrives in Ireland, 144
  come to teach the art of mining, 145
  commission by Henry VIII to recruit miners, 143
  discharged from the mines, 153
  doubts competence of Garret Harman, 143
  relegated to second in command, 145

Halliwell, James, 165
Harman, Garret, 99, 143, 144, 145, 148
Harrison, John, 140
Hartwell, Robert, 194
Hatton, Edward, 194
Herbert, Sir William (first earl of Pembroke)
  altercation at the Bristol Mint, 106
  attempts to bully Robert Recorde in London, 108
  destroys shrine at Penrhys in the Rhondda, 158
  first confrontation with Robert Recorde, 98
  gains favour with Queen Mary, 158
  his ennoblement, 158
  his later life and death, 193
  his slaying of rebels, 105
  sues Robert Recorde for libel, 181
Holinshed, Raphael, 131, 191

Indigenous sources of silver, 143

Jerome, William, 90

King Alfred, 80
Kyngston, John, 129, 182

Lathebury, Stephen, 102
Leland, John, 77, 78, 79, 131
Little England beyond Wales, 5
Little Moreton Hall, 195
Lollards, 22
Lord Protector *see* Seymour, Sir Edward
Lougher, Robert, 133, 193

Markham, Sir John, 88, 132
Mary Tudor (catholic queen of England), 155
Master and scholar
  addition of money, 47
  cost of buying a horse, 49
  identity of the master, 39
  identity of the scholar, 39
  multiplying simple numbers, 44
  number of bricks to build twelve walls, 49
Mellis, John, 199
Metal type, 58, 59
Moreton, William, 195

Mounte, Christopher, 56
Mune, John, 102
Muscovy Company
    founding, 109
    loss of ships and crews, 115
    return of Richard Chancellor, 115

Northumberland, duke of *see* Dudley, Sir John
Norton, Robert, 194
Norwood, Richard, 194

Ohthere of Halgoland, 113
Old English Manuscript No. 138, 79
Owen, George
    the Welsh/English language divide, 6, 9
Oxford University
    ceremony for incepting bachelors, 20
    commons and battelling, 21, 47
    disputations, 20
    hostel accommodation, 17
    the quadrivium, 20
    the schools, 18
    the trivium, 18

Pacioli, Luca, 168
Partridge, Nicholas, 56
Paul's Walk, 53
Pembroke, earl of *see* Herbert, Sir William
Pembrokeshire, 5
Pentyrch iron mine, 98
Pinteado, Anthony Anes, 112
Pirry, Martin, 145
Pitts, John, 3, 91
Pole, Reginald (Cardinal), 156, 158
Powell, Edward, 91
Print run for textbooks, 59
Printing ink, 58
Printing press, 58

Recorde, Joan (*née* Ysteven), 8
Recorde, Richard (brother)
    alchemical studies, 103
    becomes mayor of Tenby, 190
    his birth, 9
    his marriage and children, 134
    in Ireland with Robert, 144
    surveyor at the Bristol mint, 102
Recorde, Robert
    a characterisation by Pitts, 3
    accused of treason by Herbert, 106
    admitted Bachelor of Arts, 20
    and abilities as a poet, 134
    and commissions by London printers, 129
    and confinement to court, 107
    and conversion to Protestantism, 23–4
    and entry to Oxford University, 17
    and exposure of the false prophet Allen, 88
    and his dispute with Gundelfinger, 145
    and his invention of the 'equals' sign, 170
    and his knowledge of metals and mining, 146
    and his knowledge of navigation, 112
    and his letter to Queen Mary, 174
    and his medical care of Edward Underhill, 156
    and his names for geometrical entities, 84
    and his possible doctorate in theology, 89
    and his possible translation of Euclid, 84
    and involvement with the Muscovy Company, 109
    and Latin, 15, 18

Recorde, Robert (CONTINUED)
   and Leland's revelation of Old
      English manuscripts, 78
   and property in Tenby, 133
   and rhyming prose, 42
   and Roger Bacon's telescope, 131
   and Stephen Gardiner's sermon, 96
   and Stephen Gardiner's trial, 110
   and teaching of medicine, 64
   and the books he planned to write,
      85–6
   and the Catholic mass, 13–15
   and the ceremony of *Vestro*
      Communi, 111
   and the contents of his library, 80
   and the practice of surgery, 72
   and the Royal College of
      Physicians, 72, 122
   and the travels of Ohthere, 114
   and the whale oil incident, 147, 154
   as comptroller of Durham House
      Mint, 100
   as undertreasurer of the Bristol
      Mint, 102
   as Inspector-General of the Dublin
      mint, 144
   as iron master at Pentyrch, 99
   as surveyor of the Irish mines and
      monies, 143
   bogus portrait in oils, 2
   brings order to mining operations,
      144
   elected fellow of All Souls, 21
   his appearance, 1
   his arrest and imprisonment, 183
   his birth date, 9, 17, 110
   his Cambridge MD, 63
   his charge of malfeasance against
      the earl of Pembroke, 178
   his childhood education, 15–16
   his childhood reading, 16
   his death, 189
   his family home in Tenby, 10
   his family name, 3
   his house in St Katherine Coleman,
      66, 75
   his last will and testament, 185
   his Oxford medical licence, 21
   his paternal grandfather, 4
   his paternal grandmother, 5
   his possible burial place, 189
   his servant John, 66, 187, 189
   leaves academia to practise
      medicine, 64
   leaves Oxford for Cambridge, 24–5
   medical studies
      and the dangers of bloodletting,
         29
      and the examination of urine,
         32
      and the taking of pulses, 30
      and the theory of humours,
         29–30
   receipt of church plate, 104
   sued for libel by the earl of
      Pembroke, 181
   summoned from Ireland to
      London, 152
   summoned to appear before the
      Privy Council, 180
   supposed portrait woodcuts, 1–2
   undergraduate studies, 18
   worried by penury, 141
Recorde, Robert (nephew)
   becomes a bailiff of Tenby, 190
   granted lease of Crown lands, 190
   has four sons and five daughters,
      190
   son and heir of Richard, 134
Recorde, Roger (grandfather), 4
Recorde, Roger (uncle), 5, 8
Recorde, Roger and Elsbeth
   their arrival in Tenby, 7
   their journey to Wales, 6

Recorde, Ros (mother)
    daughter of Thomas ap John ap Sion, 9
    née Johns, of Machynlleth, 8
    Ros merch Thomas ap John, 9
Recorde, Thomas (father), 5, 8
    becomes Mayor of Tenby, 15
    his death, 103
    his first marriage, 8
    his second marriage, 8
Recorde, William (uncle), 5, 8
Reparasso, John de (Frenchman), 154
Robyns, John, 22, 190
Ryce, William, 174

St John's Croft, Tenby, 190
St Paul's Churchyard, 53
St Paul's Cross, 53
Sawtt, Elsbeth (grandmother), 4
Schmalkaldic League, 93
Seven liberal arts, 20
Seymour, Sir Edward (duke of Somerset), 84, 87, 107, 144
Sharington, Sir William, 97, 98, 100, 192
Sign of the Brazen Serpent, 54
Sir Benfro, 5
Smith, John, 102
Somerset, duke of *see* Seymour, Sir Edward
Straits of Magellan, 113

Talbot, Robert, 78, 79, 191
Tenby, 7–8
    Church of St Mary the Virgin, 13
    St Julian's chapel, 13
    the harbour, 11
    the market, 11
The Castle of Knowledge
    a treatise on cosmography and the celestial sphere, 111
    circumference of the earth, 120
    dedication to Queen Mary, 157
    first English language exposition of the Copernican thesis, 126
    fundamentals of Ptolemaic astronomy, 124
    intended for Muscovy Company navigators, 109
    latitude of Wardhouse (Vardo), 120
    the celestial spheres, 118
    the four books or parts, 117
    the four elements, earth, water, air and fire, 117
The Chronicles of England and France, 129
The false portrait of Robert Recorde
    displayed at the Ashmolean Museum, 201
    gifted to Cambridge University, 201
    no. 41 in the Old Ashmolean Series of postcards, 201
    opinion of National Portrait Gallery on its authenticity, 201
    opinion of the Fitzwilliam Museum on its authenticity, 202
    reproduced in Gunter's *Early Science at Oxford*, 201
    staggering coincidence of its discovery, 201
    supposed to be Recorde's likeness, 201
    the sculpted medallion in Tenby church, 203
The Gate of Knowledge, 85
The Grounde of Artes
    arithmetic with counters, 50
    arithmetic with the pen, 39, 44
    art of numbering on the hand, 51
    genesis, 33
    its phenomenal success, 193
    reason for writing and teaching methods, 35
    title page woodcut and VDME, 92

The Pathway to Knowledge
   criticised as less rigorous than Euclid, 81
   dedication to Edward VI, 84
   new English words for geometrical entities, 84
   simple methods of dividing an arc, 82
   the arguments of the four books, 85
The Treasure of Knowledge, 86
The Urinal of Physick
   dedication to the Wardens and Company of Surgeons in London, 72
   ethos of Tudor medicine, 68
   judical of urine, 69
   medicinal and curative properties of urine, 70
   subsequent editions, 195
The Whetstone of Witte
   algebraic notation, 167
   continuation of first treatise on arithmetic, 160
   extraordinary ending to the textbook, 171
   first English work on Algebra, 165
   invention of the 'equals' sign, 170
   strange dedication to the Muscovy Company, 182
   supposed pun on the title, 161
   symbols for plus and minus, 170
   voyage of Othere, 113
   zenzizenzizenzike, 168
Tudor dynasty, 5
Tudor physicians, 67

Underhill, Edward
   his arrest of Allen's associates, 88
   his arrest of the false prophet Allen, 87
   his later life and death, 192
   ill in prison with a burning ague, 156
   imprisonment in the reign of Queen Mary, 155
   the hot gospeller, 89
Urinal, as a glass vessel, 68

Visitation to Wales, 4

Walker, John, 102
Welsh emancipation, 25
Welsh patronymic naming system, 9
Whalley, Richard
   acquaintance with Richard Eden, 132
   his death, 191
   sons Thomas and Hugh, 34
   wives and children, 33, 37
Willoughby, Sir Hugh, 109, 114, 120
Willsford, Thomas, 194
Wolfe family members, 75–7
Wolfe, Joan, 77
Wolfe, Reyner
   and *Holinshed's Chronicle*, 131
   appointed King's printer, 60
   Brazen Serpent printing house, 57
   death of first wife, 56
   his death, 191
   his second marriage, 75
   imported books packed in barrels, 55
   intervention of Anne Boleyn, 57
   patent of denization, 55
   provides bond for Robert Recorde, 180
   removes bones of the dead from charnel house, 77
   spelling of first name, 55
   visits to Frankfurt am Main book fair, 55
Wood, Anthony à, 3, 129
Wycliffe, John, 22
Wygmore, Roger, 102, 181

Ysteven, Thomas, 8